NCRP REPORT No. 51

RADIATION PROTECTION DESIGN GUIDELINES FOR 0.1–100 MeV PARTICLE ACCELERATOR FACILITIES

Recommendations of the
NATIONAL COUNCIL ON RADIATION
PROTECTION AND MEASUREMENTS

Issued March 1, 1977
First Reprinting October 1, 1979

National Council on Radiation Protection and Measurements
7910 WOODMONT AVENUE / WASHINGTON, D.C. 20014

Copyright © National Council on Radiation
Protection and Measurements 1977

All rights reserved. This publication is protected by copyright. No part of this publication may be reproduced in any form or by any means, including photocopying, or utilized by any information storage and retrieval system without written permission from the copyright owner, except for brief quotation in critical articles or reviews.

Library of Congress Catalog Card Number 76-52067
International Standard Book Number 0-913392-33-2

Preface

This report of the National Council on Radiation Protection and Measurements (NCRP) is concerned with radiations produced by accelerators of charged particles having energies from 0.1 to 100 MeV. The material in this report includes recommendations concerning structural shielding and details of accelerator-facility design as they pertain to radiation protection.

The scientific committee responsible for the preparation of this report was charged with preparing a guide to good practice in radiation protection for all types of particle accelerators, taking into full consideration their broad application in research, medicine, and industry. In carrying out this objective, the committee has endeavored to organize into a single report the recommendations and guidelines for the many accelerator designs, performance ratings, and applications, without resorting to over-generalization or undue conservatism.

There is some overlap of this report with the coverage of other NCRP reports, but an attempt has been made to limit duplication of material except where it is justified for the sake of continuity, or because of the need to complement the coverage of the following existing NCRP reports, or to update their information and recommendations:

No. 14 *Protection Against Betatron-Synchrotron Radiations Up to 100 Million Electron Volts* (NCRP, 1954a) [Superseded by the present report];

No. 31 *Shielding for High-Energy Electron Accelerator Installations* (NCRP, 1964b) [Superseded by the present report];

No. 34 *Medical X-Ray and Gamma-Ray Protection for Energies up to 10 MeV — Structural Shielding Design and Evaluation* (NCRP, 1970a) [Superseded by NCRP Report No. 49, *Structural Shielding Design and Evaluation for Medical Use of X Rays and Gamma Rays of Energies Up to 10 MeV* (NCRP, 1976)];

No. 38 *Protection Against Neutron Radiation* (NCRP, 1971a).

The recommendations in this report provide basic standards for use in the preparation of regulatory protection codes, but they are not written for literal adoption as legal regulations. The radiation sources discussed herein may be subject to regulation by federal, state, or local governmental agencies. Such regulations may involve registration, licensing, or compliance with specific rules.

The present report was prepared by Scientific Committee 22 on Radiation Shielding for Particle Accelerators. Serving on the Committee during the preparation of this report were:

E. ALFRED BURRILL, *Chairman*

J. ROBERT BEYSTER CLARENCE J. KARZMARK
GORDON L. BROWNELL WILLIAM E. KREGER
ARTHUR B. CHILTON JAMES M. WYCKOFF
JACOB HAIMSON

DAVID GROCE (Consultant)

The Council wishes to express its appreciation to the members and consultant of the Committee for the time and effort devoted to the preparation of this report.

LAURISTON S. TAYLOR
President, NCRP

Bethesda, Maryland
October 15, 1976

Contents

Preface	iii
1. Introduction	1
1.1 Purpose	1
1.2 Scope	1
1.3 General Considerations	3
2. Facility Considerations	7
2.1 Siting and Layout	7
2.2 Radiation Safety Systems	11
2.3 Special Problems	19
3. Sources of Radiation	25
3.1 General	25
3.2 Accelerated Charged Particles	27
3.3 Penetrating Radiations	29
3.4 Other Sources of Radiation	33
4. Radiation Shielding	37
4.1 Shielding Materials	37
4.2 Parameters of Shielding Calculations	42
4.3 Methods of Calculating Shielding Thickness	49
4.4 Apertures	60
4.5 Special Shielding Problems	68
APPENDIX A. Definitions	73
A-1 Definition of Terms	73
A-2 Definition of Symbols	83
APPENDIX B. Data Pertaining to Dose Limits and Radiation Effects on Materials	85
B-1 Dose-Limiting Recommendations	85
B-2 Quality Factors for X Rays and Electrons	86
B-3 Mean Quality Factors, $\bar{Q}$, and Fluence Rates per Unit Dose-Equivalent Rate for Monoenergetic Neutrons	87
B-4 Thresholds for Radiation Damage to Selected Materials and Systems	88
B-5 Typical Workload, W (for Busy Radiotherapy Installations Only)	90
B-6 Area-Occupancy Factor, T	91
APPENDIX C. Conversion Factors and Equivalents	92

APPENDIX D. Charged Particles 93
  D-1 Range of Monoenergetic Electrons 93
  D-2 Fraction of Electrons Backscattered 94
  D-3 Range of Protons 95

APPENDIX E. X Rays and Gamma Rays 96
  E-1 X-Ray Emission Rates from High-Z Targets 96
  E-2 Angular Distribution of Emitted X Rays from High-Z Targets 98
  E-3 X-Ray Emission Rates from Electron Impingement on Thick Targets of Low-Z Material 99
  E-4 X-Ray Emission Rates from Electron Backstreaming in Direct Proton Accelerators 100
  E-5 Radiations from Two-Stage Tandem Accelerators 101
  E-6 Equivalent Incident Electron Energies 102
  E-7 Broad-Beam Transmission Through Concrete of X Rays Produced by 0.1–0.4-MeV Electrons 103
  E-8 Broad-Beam Transmission Through Concrete of X Rays Produced by 0.5–176-MeV Electrons 104
  E-9 Broad-Beam Transmission Through Steel of X Rays Produced by 1–31-MeV Electrons 105
  E-10 Broad-Beam Transmission Through Lead of X Rays Produced by 0.1–0.4-MeV Electrons 106
  E-11 Broad-Beam Transmission Through Lead of X Rays Produced by 0.5–86-MeV Electrons 107
  E-12 Dose-Equivalent Index Tenth-Value Layers for Broad-Beam X Rays in Concrete 108
  E-13 Dose-Equivalent Index Tenth-Value Layers for Broad-Beam X Rays in Steel 109
  E-14 Dose-Equivalent Index Tenth-Value Layers for Broad-Beam X Rays in Lead 110
  E-15 Reflection Coefficients for Monoenergetic X Rays in Concrete, Iron and Lead 111

APPENDIX F. Neutrons 112
  F-1 Thick-Target Neutron Fluence Rates for (p, n) Reactions 112
  F-2 Thick-Target Neutron Yields for (d, n) Reactions 114
  F-3 Thick-Target Neutron Yields for (γ, n) Reactions 115
  F-4 Typical Angular Distributions in Neutron Yield Ratios from Several Neutron-Producing Reactions 116
  F-5 Maximum Energies of Neutrons Produced from Proton and Deuteron Reactions 117
  F-6 Dose-Equivalent Index Transmission Through Concrete of Monoenergetic Neutrons 118

CONTENTS / vii

F-7   Accelerator Conditions for Generating Neutron Spectra Referred to in Appendices F-8 and F-9 120
F-8   Dose-Equivalent Index Transmission Through Concrete of Neutrons from (γ, n) and (γ, fn) Reactions ... 121
F-9   Dose-Equivalent Index Transmission Through Concrete of Neutrons from Ion-Induced Reactions 122
F-10 Dose-Equivalent Index Tenth-Value Layers in Concrete for Monoenergetic Neutrons 124
F-11 Thermal-Neutron Transmission Through Mazes and Ducts 125
F-12 Reflection Coefficients for Monoenergetic Neutrons in Concrete, Iron and Lead 126

APPENDIX G. **Radioactivity** 127
G-1   Radioactivity-Producing Nuclear Reactions 127
G-2   Induced Radioactivity in Cyclotrons 128
G-3   Gamma-Radiation Dose-Equivalent Index Rate Following Cyclotron Shutdown 129

APPENDIX H. **Shielding Materials** 130
H-1   Densities of X-Ray Shielding Materials 130
H-2   Neutron Shielding Materials: Elemental Composition of Concretes 131
H-3   Neutron Shielding Materials: Effects of Water Content and Elemental Composition on Neutron-Transmission Properties of Concrete 132
H-4   Neutron Shielding Materials: Miscellaneous Materials 133

APPENDIX I. **Ozone and Other Noxious Gases** 134
I-1   Production of Ozone by External Electron Beams 134
I-2   Threshold Limit Values for Ozone and Certain Oxides of Nitrogen in Workroom Air 136

References 137
The NCRP 145
NCRP Reports 151
Index 155

1. Introduction

1.1 Purpose

The purpose of this report is to provide design guidelines for radiation protection in particle-accelerator facilities, and to describe one or more methods by which this protection may be achieved. The recommendations herein may well be modified in unusual circumstances upon the professional advice of experts with recognized competence in radiation protection for particle accelerators.

This report is directed mainly to designers of accelerator facilities, and is written from an engineering point of view. It is believed that health and radiological physicists, research scientists, project engineers, technical administrators, and similar specialists will also find the information useful.

1.2 Scope

1.2.1 *Accelerators*

This report is concerned with accelerators producing particles having energies below 100 MeV (see Tables 1 and 2), with exceptions and qualifications as itemized below:

a. *X-ray generators:* All x-ray generators above 0.1 MV are considered to be within the scope of this report. However, NCRP Report No. 49 (NCRP, 1976) is concerned with medical x-ray protection for energies up to 10 MeV. Structural shielding recommendations and x-ray transmission data given in NCRP Report No. 49 are applicable as well to non-medical applications of low-energy x-ray generators, particularly for energies below 0.5 MeV. The present report includes x-ray transmission data and shielding-calculation bases for these very low energies, but is primarily concerned with x-ray generators with accelerated electron energies above 0.5 MeV.

b. *Neutron generators:* All particle accelerators that can produce neutrons are considered to be within the scope of this report. How-

TABLE 1—*Types of particle accelerators*

Direct (Potential-Drop)
(Single-stage for acceleration of either ions or electrons)
(Two-stage [tandem] for acceleration of ions)
 1. *Electrostatic high-voltage generators:*
     a. belt-charging system (e.g., Van de Graaff, Pelletron),
     b. rotating-cylinder charging system.
 2. *High-voltage transformers:*
     a. transformer-rectifier set,
     b. voltage-multiplying system (e.g., Cockcroft-Walton, Dynamitron),
     c. cascaded transformer system (e.g., Insulating-Core Transformer).

Indirect (Cyclic)
 1. *Linear Beam Trajectory:*
     a. ion linear accelerator,
     b. electron linear accelerator.
 2. *Circular or Spiral Beam Trajectory:*
     a. cyclotron   (ions only),
     b. synchrotron   (ions or electrons),
     c. betatron   (electrons only),
     d. microtron   (electrons only).

TABLE 2—*Applications of particle accelerators*

	Electron	X Ray	Ion	Neutron
Diagnostic Radiology		*		
Radiotherapy	*	*	*	*
Industrial radiography		*		*
Analysis of materials, e.g.,				
activation analysis,			*	*
microscopy, electron or ion	*		*	
x-ray fluorescence analysis	*	*	*	
Ion implantation, polishing			*	
Radioisotope production			*	*
Research and training, e.g.,	*	*	*	*
nuclear-structure physics				
neutron physics				
atomic and solid-state physics				
biology, chemistry				
radiation effects on materials				

ever, a forthcoming NCRP Report (NCRP, 1977) is concerned with protection for small neutron generators, i.e. deuteron accelerators below 0.4 MeV in particle energy. The present report is therefore primarily concerned with neutron-producing accelerators that accelerate charged particles to energies above 0.4 MeV.

c. *Electron accelerators:* All electron accelerators with *externally* applied beams as low as 0.1 MeV in particle energy are included, whether used in medical or non-medical application.

Accelerators that produce particles with energies above 100 MeV are considered to be outside of the scope of this report. Thus, meson production with its attendant problems is not discussed.

1.2.2 Approach to the Problem

In order to develop recommendations for the protection of personnel and the general public from the radiations produced by particle accelerators, it is important to understand the relevant characteristics of radiations that *can* be produced by *all* types of accelerators in *all* presently recognized applications. Wherever possible in this report, experimental data have been used, either as a primary basis for shielding calculations or as a check on theoretical approaches. In those cases where exact solutions to shielding problems cannot be derived without extensive computations, conservative approximations are recommended that are based on experience in the field.

Recommendations contained in this report are expressed in terms of *shall* and *should*. The use of these terms is defined as follows:
 (1) *Shall* indicates a recommendation that is necessary to meet the currently accepted standards of radiation protection.
 (2) *Should* indicates a recommendation that is to be applied when practicable.

Other terms and symbols used in this report are defined in Appendix A.

1.3 General Considerations

1.3.1 Exposure of Individuals

Reduction of radiation exposure to an individual from an external source of radiation may be achieved by any one or any reasonable combination of: (a) increasing the distance from the source of radiation; (b) placing physical shielding barriers between the individual and the source; (c) reducing the duration of the exposure; or (d) reducing the radiation energy and/or emission rate from the source of radiation.

Recommendations on controlling exposure of individuals are presented in terms of *maximum permissible dose equivalents* (H_M) for occupational workers and *dose limits* for the general public set out in NCRP Report No. 39, *Basic Radiation Protection Criteria* (NCRP, 1971b). The recommended values of H_M and dose limits, as well as associated considerations, are summarized in Appendix B of this report. They constitute the values of H_M and dose limits used in this report. *Dose equivalent* is defined in Appendix A.

The magnitude of the exposure of individuals that *should* be permitted, whether for occupational workers or for the general public, has been the subject of serious evaluation for years. The primary objective in establishing H_M values for occupational exposure is to keep the exposure of the radiation worker well below a level at which adverse effects are likely to be observed during one's lifetime. Another objective is to minimize the incidence of genetic effects for the population as a whole. It must be emphasized that the risk to individuals exposed to the H_M or to the dose limits for the population is considered to be very small; however, risk increases with increasing dose.

For this reason, it is desirable to keep radiation exposures as low as practicable, by using H_M values and dose limits as an upper limit in accelerator-facility design rather than as a prescribed design basis.

It should be noted that federal, state, and local governments may impose H_M values and dose limits that differ from NCRP recommendations.

1.3.2 *Design Basis of Accelerator Facility*

An accelerator-facility design *shall* be based on the maximum radiation emission rate of the accelerator, i.e. with whatever achievable combination of radiation species, energy, and beam current that requires the most shielding. The possibility of later operating at higher energies and/or beam currents *should* be anticipated in the original facility design. After the facility has been completed, later shielding additions can be expensive and time-consuming, as well as utilizing space that is often committed. Hence, adequate structural shielding *should* be properly designed and installed in the original construction process.

The facility designer can often obtain from the accelerator manufacturer necessary information concerning the emission rates of radiations from the specific accelerator under consideration. In the absence of sufficient information, however, the accelerator-facility designer *shall* assume that the radiation characteristics are the same as for equivalent or larger accelerators operating according to the same principles of operation. Section 2 includes procedures for estimating radiation emission rates from typical accelerators.

1.3.3 *Radiation Protection Surveillance*

It is recommended that the shielding be designed with the counsel of a qualified expert on radiation shielding (see Appendix A), to ensure ample radiation protection at minimal cost. This individual *should* be provided with all pertinent information regarding the proposed accelerator and its immediate and projected uses, type of building construction, and occupancy of nearby areas. This expert *should* be consulted during the early planning stages. The shielding requirements may affect the choice of location of accelerator facilities and the type of building construction; on the other hand, the radiation-safety considerations may affect certain details of the design of the accelerator.

During the construction of the accelerator facility, the qualified expert on radiation shielding *should* inspect the progress of the work at appropriate times to ensure that the radiation-protection aspects of the construction are in accordance with the design specifications.

When the facility construction has been completed and/or the accelerator is ready for initial operation, the accelerator *should* be brought gradually up to its maximum radiation emission rate, with the attendance of a qualified health physicist or qualified radiological physicist (not *necessarily* the same individual as the qualified expert on radiation shielding, see Appendix A), who *shall* be engaged to make a radiation survey of the entire installation, to ensure that H_M values and dose limits will not be exceeded. If the survey indicates that the applicable H_M or dose limit could be exceeded, the qualified expert(s) *shall* recommend appropriate corrective measures.

If at a later time, the operating parameters of the accelerator are changed so as to increase the radiation emission rate, a qualified health physicist or qualified radiological physicist *shall* be engaged to resurvey the entire installation to ensure that the shielding is still adequate. For accelerators with beam-transport systems, resurveys *shall* be made whenever significant changes are made in beam steering, collimation, or local and temporary shielding.

1.3.4 *Dose Equivalent Index*

For purposes of specifying ambient radiation levels for use in radiation protection, the International Commission on Radiation Units and Measurements (ICRU) has defined the quantity dose equivalent index. This is the *maximum* dose equivalent within a particular mass, namely a 30-cm diameter sphere, centered at the point of interest and consisting of material equivalent to soft tissue with unit density (ICRU, 1971a). In order to express radiation levels in terms of dose equivalent index, H_1, calculations for, or measurements in, a 30-cm tissue-equivalent sphere must be made.

A question arises as to how one obtains the neutron dose equivalent index from the neutron fluence and data now given for normally incident neutrons in terms of the maximum dose equivalent in either a 30-cm thick slab or a 30-cm diameter cylinder of tissue-equivalent material. Calculations of the neutron fluence rate per unit maximum dose equivalent rate have been performed for a cylindrical geometry (NCRP, 1971a; Auxier et al., 1968) and for a slab geometry (NCRP, 1957; Irving et al., 1967; Zerby and Kinney, 1965; Alsmiller et al., 1970; Wright et al., 1969). The results of most of these calculations agree within about 20 percent [see summary plot of most of these results in ICRP Publication 21 (ICRP, 1971)]. The calculations of Auxier et al. (1968), as well as those of NCRP (1971a), incorporate new cross-section data, include more reactions than had previously been used, take into account a first approximation correction for scattering, and also include a more detailed consideration of the distribution of dose with collision stopping power. On the other hand, the spatial resolution in the calculations is not as good for the cylinder as for the slab.

Thus, with the information currently available, and within the accuracy described, the quotients of the neutron fluence rate per unit maximum dose equivalent rate for the slab and cylindrical geometries can be considered to be the same. On this basis it is suggested that the same values also be used for the spherical geometry of similar dimensions. While further verification will be necessary, with this suggestion the values of the maximum dose equivalent will be considered to be the same for slab, cylindrical, and spherical geometries.

2. Facility Considerations

2.1 Siting and Layout

2.1.1 *Typical Sites*

The cost of radiation protection for particle accelerators depends to a large extent on the occupancy of surrounding areas, as well as on the radiation emission rate expected from the specific accelerator. Ideally, the accelerator facility should be located as far as practicable from occupied areas. However, considerations of convenience, availability of existing buildings, association with industrial processes or medical facilities often override the protection advantages of a remotely located installation.

A *separate building* or a *wing* may often be the best site for an accelerator facility, because its design can be specifically developed for the purpose, and subsequent changes can be effected with greater ease. Adequate footings can be readily provided for the weight of the accelerator as well as its associated equipment and the radiation shielding. In some research-oriented facilities, anticipation of a possible future requirement for additional radiation shielding can be incorporated into the initial design. If the terrain permits, an underground installation significantly reduces the shielding problem. Alternatively, earth can be built up around the structural walls, as supplementary shielding to a grade-level structure. Excavation into a hillside is another method of utilizing the shielding advantages of the terrain. If land is inexpensively available, fences can be erected at some distance around the facility to exclude personnel and the general public from areas where the H_M or dose-limit values may be exceeded.

Modification of an existing structure can be undertaken, provided that the building design can support the additional weight of the accelerator, its associated equipment, and the necessary radiation shielding. Horizontally oriented accelerators are often installed in such modified facilities. Headroom limitations tend to inhibit the installation of vertically mounted machines, unless space on the floor(s) above is available. Support for lifting equipment must be provided for installing and servicing the accelerator. The major difficulty in designing an accelerator facility into an existing structure involves the radiation shielding. Usually, such buildings are occupied on all sides and occasionally above (and/or below) the planned facility. Shielding walls can be supported if there are adequate footings, and the walls are supported directly by these footings. However, special structures may have to be provided to support the shielding required for the ceiling. Space is often at a premium in existing buildings, and radiation shielding barriers of ordinary structural materials may be too bulky to provide adequate room in the allotted floor space for the accelerator and its use. In such instances, special shielding materials with high radiation-attenuation characteristics may have to be utilized. These are generaly more expensive and often require special attention in their erection.

Industrial process lines can occasionally be best served by an accelerator incorporated in the line of material flow. Such installations may be classified as noncontrolled areas and, as such, must be carefully designed, not only to accommodate the material to be processed, but especially to protect persons in the immediate vicinity from *nonoccupational* exposure to the radiations produced by the accelerator.

Portable or mobile accelerators, such as certain types of x-ray generators and neutron generators, are occasionally put to use temporarily in buildings or in the field, where permanent radiation shielding is not available. In such instances, special attention must be paid to the protection of the general public, as well as the radiation worker, by the erection of temporary shielding or by providing an adequate exclusion area around the accelerator while it is in use.

Architectural and engineering problems in siting, other than those mentioned above, must also be considered, but they are beyond the scope of this report. For example, large and complex accelerators require stable soil conditions so that accurate beam alignment can be maintained. Underground facilities must be protected against water seepage.

2.1.2 *Space Requirements*

For an accelerator facility to be used effectively, adequate space must be provided for all anticipated uses. Experience has shown that radiation-protection designs and procedures tend to be inadequate in cramped quarters.

The accelerator room must include adequate space for servicing. Many direct accelerators require space for withdrawing the pressure vessel from the accelerating column. At least initially, the access to any accelerator room must be large enough to permit moving in the largest single component of the machine. In some cases, an aperture is left unfilled in the shielding wall until after the accelerator has been moved into the room. This aperture must be filled in a manner that not only provides shielding equivalent to the remainder of the wall, but also avoids cracks through which radiation can stream. In block type construction, successive layers *should* overlap in such a fashion that joints in layers do not coincide. Space must be allowed for those items of auxiliary apparatus that must be located close to the accelerator, such as vacuum pumps and refrigeration equipment. In some installations, a pit or a trench is provided below the basic machine to accommodate these components and to conserve floor space. The accelerator room usually requires radiation shielding, even when separated from the irradiation room or experimental area, except in those special cases in which adequate shielding is incorporated in the accelerator itself.

For many kinds of low-energy machines, the *control room* need be nothing more than an alcove outside or even within the accelerator room. Compact neutron generators, x-ray generators, electron-beam generators, small accelerators for teaching, ion implantation, atomic research, etc., have relatively simple controls that are contained in a desk-mounted console, or in a self-standing cabinet. The larger the accelerator, the more complex the control system becomes. The controls and electronics for some experimental programs may desirably be located close to the accelerator controls. The space requirements for the control room can thus expand according to specific needs. The control room is usually located nearby (but not necessarily contiguous to) the accelerator, preferably in a direction away from that of the highest radiation emission rate.

The irradiation room or experimental area(s) requires adequate space for any necessary air-removal ducts, material transport, patient maneuvering, auxiliary apparatus, and maintenance activities. Experimental areas of research accelerators unfortunately are rarely designed with adequate space for an expanding investigative program. For many types of research, the auxiliary equipment tends to become almost as bulky as the accelerator itself. Rooms for x-ray or neutron research purposes are sometimes designed with a large free volume to minimize the effects of neutron or x-ray scattering or reflection[1] from the walls, ceiling, and floor; the latter is usually

---

[1] In describing the backscattering action of a barrier upon incident radiation, the term "reflection" will be used herein to distinguish the backscattering process as a macroscopic concept proceeding in a specific angle of interest, from the atomic process called "scattering." This prevents confusion, e.g., when talking of a single reflection by a barrier, in which the atomic processes involved actually include multiple scattering.

accomplished by means of a pit below the x-ray or neutron-producing target that may be covered by a low-atomic-number (low-Z) and/or expanded-metal false deck.

Beam dumps for x rays or neutrons are frequently incorporated into the design of experimental areas for high-energy, high-power accelerators. Their purpose is to accept the radiation beam into a shielded cavity, of such dimensions and materials that the reflection of the radiation is significantly reduced. These beam dumps are located in line with the accelerated-beam axis in each area to which the beam can be magnetically switched. In some beam-dump designs, the lining of the cavity is chosen for its low radiation-reflection characteristics. The entrance aperture into the cavity should be only large enough to accept the major components of the x-ray or neutron beam. In the case of accelerators performing with particle energies of tens of MeV, the angle enclosing virtually all of the radiation from the target is a few degrees. The depth of the cavity should be greater than the diameter of the aperture, to reduce the solid angle of back reflection. The radiation shielding surrounding the cavity should, of course, provide adequate protection for personnel in the area beyond and to the side of the beam dump.

Other areas that are contiguous to the radiation source(s) may have to be used for such purposes as low-level counting rooms, photographic or radiographic darkrooms, shop facilities, or laboratory offices. All of these areas must be provided with adequate radiation protection for personnel. The counting rooms and darkroom may require additional shielding because of the nature of the work or the radiation sensitivity of the equipment or photographic materials.

Architectural considerations that may involve several areas of an accelerator facility include interconnecting cable ways or ducts, ventilation and/or air-conditioning ducts, trenches for piping, conduits, or auxiliary apparatus, and pits for the storage of radioactive targets. Possible radiation damage to certain components and materials *should* be considered (see Section 2.3.3). Research accelerators may require special materials-handling equipment such as traveling-crane hoists, elevators, or (occasionally) remotely operated equipment. The simpler, single-purpose accelerators can often be installed without such extensive design considerations, because their cables and water lines are small in number and are usually supplied by the manufacturer as part of the equipment.

2.2 Radiation Safety Systems

2.2.1 *General*

This section deals with auxiliary requirements for radiation protection that may affect the design of an accelerator facility. The recommendations given below are not meant to preclude alternative methods of achieving radiation-protection objectives, and they may be modified upon the advice of a qualified expert.

The purpose of a radiation safety system is to assist in assuring that proper operating procedures are followed. The success of such a system depends inevitably on the people who are to be associated with it. Not only *shall* they be familiar with the system and its procedures; they *shall* respect the system and adhere to its procedures.

Persons who are to enter the controlled area *shall* be instructed, commensurate with their activity in the area, in the safety procedures and precautions to be applied to avoid or minimize exposure. Instruction *should* include explanation of the location, operation, and significance of warning signals, emergency switches, and interlock systems.

It is impractical to design a radiation safety system that is fully protective for the individual who *willfully* exposes himself to high radiation fields. On the other hand, a system that cannot inherently provide protection under a plausible combination of circumstances must be considered inadequate — and, indeed, it may be worse than no system at all, since it may give an illusion of protection to those working in the area.

One essential criterion for an adequate radiation-safety system is that responsibility for safe operation must reside *at any given time* with a single designated and qualified individual.

2.2.2 *Interlocks*

Personnel entrances into any high-radiation area or exclusion area *shall* be provided with either a door with shielding equivalent to that required of the surrounding walls or a physical barrier (in the form of a door, gate, or chain) at the entrance to a radiation-protective labyrinth or maze. In either case, the radiation level outside the barrier *shall* be no greater than that from the adjacent portions of the shield. All personnel-access barriers *shall* be equipped with interlock switches that cause the production of radiation by the accelerator to stop if the access barrier is opened or removed. Interlock switches *shall not* be used as a means for routinely shutting down an accelerator. It *shall not* be possible to start up an accelerator except from the control center.

The number of personnel-access portals into high-radiation or exclusion areas *should* be kept to a minimum. Where more than one personnel access is available, all such accesses *shall* be equipped with interlock switches, so that access by personnel *shall not* be possible while radiation is being produced. Those accesses that are not directly visible from the control center *shall* be equipped with additional locking arrangements, interlocked with the control system.

All interlock systems *shall* be designed to be as fail-safe as possible so that, in the event of mechanical and/or electrical failure of any components of the interlock, it *shall not*, as far as possible, cause radiation to be produced. Mechanical and electrical components *should* therefore be rugged, reliable, and tamperproof in design. Redundancy in frequently used interlock systems is desirable. In cases where extensive interlock systems are required, more than one energizing loop *should* be incorporated into the control system. Wherever appropriate, interlock components *should* be resistant to radiation-induced damage or, alternatively, *should* be protected from high radiation levels. All interlocks *shall* be inspected and tested periodically to insure that they are functioning as designed.

Interlock-system bypasses *should* be kept at a minimum. Their use, for servicing or observation of the accelerator, *shall* be permitted only under carefully controlled and monitored conditions. The individual that is designated to be responsible for safe operation *shall* be the only individual who can authorize "bypassing", and he *shall* be responsible for the removal of all bypasses before normal operation is resumed.

Emergency switches to stop the production of radiation *shall* be placed conspicuously in high-radiation and exclusion areas, so that personnel within such areas can have ready access to them in the event that they are inadvertently caught within the area upon the occasion of radiation being produced. The emergency switches *shall* be part of the interlock system. All switches *shall* be clearly and conspicuously marked as to their intended function.

Interlocks *shall* be provided to protect personnel from electrical hazards such as high-voltage power supplies and linear-accelerator (linac) modulators.

Movable radiation shielding, that supplements the accelerator-facility shielding in controlling radiation levels, *shall* be provided with interlock switches so that radiation cannot be produced in the event that the movable shielding is not in its proper position.

Certain accelerators for radiotherapy can alternatively deliver x rays or electrons to a patient. These *shall* be provided with a suitable interlock system to prevent inadvertent exposure of the patient to electrons when x-ray exposure is intended, and to cause the electron-beam current to be limited to values consistent with electron-beam therapy (as contrasted to the much higher values that are normally used for x-ray therapy).

2.2.3 *Warning Systems*

All locations designated as radiation areas, high-radiation areas, and exclusion areas – as well as the entrances to such locations – *shall* be equipped with easily observable flashing or rotating lights that operate automatically when radiation is being produced or even when the accelerator controls are set to produce radiation. Signs explaining the significance of the signal *shall* be posted in the locations, adjacent to the lights. An accelerator is considered to be a potential source of radiation when all necessary controls have been activated with the intent to produce radiation, *even if* the voltage or current or radio-frequency controls (or similar adjustable accelerator parameters) have not been sufficiently advanced to produce radiation.

The radiation-warning lights *shall* be designed into a fail-safe circuit that is tied into the interlock system, so that radiation cannot be produced if any of the warning lights have burned out.

In a large facility or in a noisy environment, an audible warning *should* also be given of the intent to produce radiation. Horns, buzzers, or public-address system speakers *should* be installed in the radiation areas. The audible signal *should* be distinctive from other warning alarms and loud enough to be heard above the ambient noise level and of sufficient duration that personnel would have adequate time either to reach the nearest emergency switch or to exit the high-radiation or exclusion area before the radiation can be turned on. In addition, proper layout and good housekeeping *should* permit fast egress from the radiation area. The audible warning *shall* be tied into the interlock system so that radiation cannot be produced until after the audible warning has ceased. It may be desirable, in complex facilities where various types of workers may be present in high-radiation areas, to cause the overhead lights to be dimmed and radiation-warning flashing or rotating lights to be activated at the same time the audible warning is sounded.

Colors of radiation barriers and outlining strips of pathways to radiation areas, and colors of radiation-warning lights *shall* be in accordance with American National Standard Z53.1-1971 (ANSI, 1971).

Warning devices, whether visible or audible, whether activated electrically or by radiation, *shall* be inspected and tested periodically to ensure that they are functioning properly as designed. Records *shall* be kept of such inspections. The length of time between inspections would depend on individual radiation conditions. In the absence of experience, annual inspections *should* be made, as a minimum requirement. Devices that are sensitive to radiation damage *should* be protected from high radiation fluences and *should* be replaced on an appropriate schedule (see Section 2.3.3).

Lights that warn of start-up conditions of an accelerator *shall* be of a different color than the radiation-warning lights, to avoid confusion concerning the status of the facility. Start-up conditions might include turning on the magnet of a cyclotron, the radio-frequency power of an electron linac, the low-voltage injector of a complex accelerator, etc. Any of these circuits may require interlocking to protect personnel, but such circuits *should* be independent of the interlock system used for radiation protection.

2.2.4 *Accelerator Operation*

The control power to the radiation-producing circuits of an accelerator *shall* be turned on only by means of a keyed switch, the key to which is under the surveillance and responsibility of the person responsible for personnel protection in the facility. The controls of low energy machines are usually so simple that the keyed switch is used to turn on *all* control power to the accelerator. In more complex machines, however, it is safe and often desirable to control independently the power to auxiliary apparatus and instrumentation, such as vacuum-pumping systems, magnets, dc power to linac modulators. *Radiation-producing circuits shall* include all circuits that, when sequentially or simultaneously activated, can create a radiation hazard; for example, the radio-frequency excitation circuits in cyclotrons, the modulator-activation circuits in linacs, the charging-belt drive circuits in Van de Graaff generators, and the high-voltage power supply circuits for the acceleration of particles.

The control system of an accelerator *shall* couple in series the facility-interlock circuits with the radiation-producing circuit so that radiation cannot be produced until the interlock system has been completely closed. This recommendation is in addition to the normally built-in protective systems of the accelerator itself that prevent improper operation of the machine in the event that one or more of its components has failed or is malfunctioning.

Recommendations for accelerator start-up procedures are beyond the scope of this report, because the details of such procedures depend on the complexity and use of the accelerator. In any case, however, a procedure *shall* be developed, with the advice of a qualified expert, to ensure that all high-radiation and exclusion areas are first cleared of personnel, prior to the emission of radiation or the intent to do so. In this connection, the person responsible for radiation protection *shall* make a personal on-the-spot survey of high-radiation and exclusion areas of the facility, to assure himself that all personnel have left. Lights or audible signal systems may be used after this inspection, but the on-the-spot check *shall not* be omitted.

After an area has been inspected and closed, an audible and/or visual signal in the area *shall* announce that a radiation hazard will exist in the area at the cessation of the signal. This precaution is necessary to warn any personnel inadvertently left behind, so that they can break the interlock circuit by an emergency switch or by opening an exit door of the area. The signal *shall* be of sufficient duration to permit this emergency action.

2.2.5 *Radiation Monitoring Systems*

It is beyond the scope of this report to go into all of the details of the appropriate instrumentation and procedures for monitoring radiation, but it is important to point out *typical* monitoring systems as components of a facility design. Reference is made to ICRU Report 20 (ICRU, 1971b) for additional information and guidance on this subject.

Area radiation monitoring systems *shall* be provided in all areas that can be occupied by personnel, where radiation levels might rise above H_M values. The monitors *shall* be continuous in operation, and their outputs *should* be displayed at the control center. If the radiation in an occupiable area rises above preset levels, an audible and/or visual warning *shall* be activated, either automatically or manually, in the area. The monitor *shall* be as fail-safe as possible, i.e. the accelerator *shall* become inoperative if the monitor fails or malfunctions.

In those installations where a mixed field of neutrons, gamma rays, and/or x rays may be generated, the monitor *shall* provide an adequate measure of the total dose-equivalent rate of the radiation fields. This may necessitate a combination of monitoring systems.

Pulsed sources of radiation impose special problems with respect to radiation monitoring. Many types of radiation monitors are duty-cycle dependent, i.e. they do not accurately indicate the time-average dose rate from pulsed radiation sources. Reference is made to ICRU Report 20 (ICRU, 1971b) for additional guidance on this subject.

An indirect measure of the emission rate from the radiation-producing target of an accelerator is sometimes used in single-purpose facilities, as a means of monitoring the radiation in the area. *Indirect* measures of radiation levels, however, *shall not* be considered adequate for personnel protection. For example, radiotherapy x-ray generators are usually equipped with a parallel-plate transmission ionization chamber placed in the useful beam, the purpose of which is to monitor the absorbed dose and/or the absorbed dose rate to a patient. These instruments usually have a preset feature to prevent overdose to the patient. A back-up dosimeter or timer *shall* be used to provide secondary turn-off capability. Neutron generators that are used expressly for activation analysis or material irradiation are often monitored by a neutron-flux-density measuring instrument located close to the target. Electron-beam generators for processing are usually monitored by the electrical parameters of the machine, such as accelerating voltage and electron-beam current. These parameters monitor the electron-beam power and, *indirectly*, the x-ray emission rate. Such radiation measuring systems *shall not* be considered as area radiation monitoring systems.

In those accelerator facilities where induced radioactivity throughout the machine may rise to high levels, an area radiation monitor can sometimes provide a useful measure of the hazard. However, the complete monitoring of induced radioactivity cannot always be successfully accomplished by the use of area radiation monitors, because the radioactivity may be concentrated in certain regions or "hot spots" of the accelerating system that can be exposed to the charged-particle beam. Wherever radioactivity is suspected, therefore, in spite of a low signal from an area radiation monitor, personnel entering the suspected area *shall* monitor the surroundings with a survey meter as they proceed into the area, and they *shall not* enter a region in which the radiation level is above prescribed H_M limits.

Visual observation of radiation areas is useful for determining whether or not personnel are located in the area, but in general this type of observation is of value only in small facilities, e.g. radiotherapy rooms, small industrial radiographic installations, low-energy research or irradiation facilities. Direct viewing can be accomplished through radiation-shielded windows, or by means of periscopic arrangements through mazes, or by observation at the personnel accesses. Closed-circuit television is occasionally used for personnel observation, especially in those installations where the TV system is also used for the direct observation of the progress of experiments or the accelerator performance in the radiation area. The installation of TV cameras must be planned with careful regard for the probable absorbed dose rate of the radiation field in which the optics and electronics of the TV system are to be located. Radiation damage can occur, in the form of progressively darkening lenses or in the malfunction or degradation of electronic components (see Section 2.3.3). If possible, the TV camera should be protected from high radiation fields by a shielded enclosure with a mirror to view the desired area, or by distance with a telephoto lens.

2.2.6 *Industrial Safety Precautions*

Procedures and precautions to ensure personnel safety in most industrial environments are well established, and this subsection merely points out certain specific industrial hazards that are often to be found in accelerator facilities.

Electric circuits and interconnections *should* be wired and installed in accordance with accepted electrical building codes.

Methods of handling, conveying, and storing materials, as well as moving heavy apparatus, *should* follow accepted industrial practice.

A supply of suitable first-aid materials and apparatus *shall* be available near the accelerator facility, including necessary apparatus for use in connection with an electrical accident.

Good housekeeping, i.e. cleanliness and orderliness, is essential in any accelerator facility, to minimize accidents, contamination, or damage.

Toxic and noxious solvents and materials are often to be found in accelerator facilities for servicing the equipment or, for example, as special gases or compounds for producing heavy ions. As a minimum, accepted industrial practice *should* be followed in the handling, storage, and disposal of such materials. In some research-accelerator installations, mercury-diffusion vacuum pumping systems are used. Care *shall* be exercised in the handling of this toxic element and its vapors.

Many direct accelerators are designed for operation in special insulating-gas atmospheres, often at high pressure. The storage and transferral of these gases, as well as their disposal, *shall* follow accepted codes of practice. The most commonly used insulating gases are mixtures of nitrogen and carbon dioxide, and sulfur hexafluoride. Inert gases such as nitrogen and carbon dioxide can be hazardous to personnel if the area is not well ventilated or if the discharged gas is not vented to the outside. For example, these gases do not support life; nitrogen, in particular, does not trigger a warning of its presence to anyone breathing it. The more complex gases mentioned above may include hazardous decomposition products after exposure to high-voltage corona or sparking.

Fire hazards are present with all kinds of electrical apparatus, and accelerators are no exception. Fire extinguishers of the appropriate type to combat electrical or solvent fires *shall* be conspicuously installed near the probable sources of trouble.

Electron-beam generators, i.e. electron accelerators whose beam is used in the atmosphere, represent an unusual type of fire hazard, arising from any one or a combination of three basic causes:

(a) *Thermal heating effects:* Electrons are capable of delivering large doses to relatively limited volumes of irradiated material because of their limited range in materials. Almost all of the energy of an absorbed electron will appear as local heating in the material. If the material is a poor conductor of heat, high internal temperatures may be attained, e.g. several hundred degrees centigrade in a few seconds of irradiation. Under unfavorable circumstances, an internal slow burning can be initiated that could later burst into an open flame.

(b) *Radiation catalysis:* A number of chemical reactions can be initiated by electron-beam irradiation. If these reactions are sufficiently exothermic, they may be self-propagating in large volumes of material and thus give rise to additional hazards from explosion or fire. In general, the absorbed dose necessary to produce radiation catalysis may be much less than that which produces severe thermal effects.

(c) *Electrical effects of irradiation:* Whenever substantial currents (greater than 100 μA) of electron beams are used, it is observed that charging and ionization effects occur in the vicinity from impingement of scattered electrons, usually leading to a cascade of tiny electrical sparks from surrounding objects. If volatile organic materials are under irradiation, the vapors from the materials may form an explosive mixture. The presence of radiation-induced sparking could provide a triggering mechanism for such reactions.

From the foregoing paragraphs, it should be obvious that a potential fire hazard exists at all times in accelerator installations. Fire-prevention authorities *should* be consulted to ensure that appropriate fire-proofing and fire-extinguishing equipment is provided.

Where volatile compounds are to be irradiated, it is further recommended that adequate ventilation be provided to prevent the accumulation of explosive concentration of vapors (see also Section 2.3.1).

Industrial electron-beam processing facilities *should* be interlocked with the flow of material to be irradiated, so that the machine can instantly be shut off in the event that there is a jam or stoppage in the material flow under the electron beam. Otherwise, the material might receive a great overdose of ionizing radiation and might burst into flame. In addition to the potential fire hazard, there is also the possibility of extensive damage to the irradiated product and to the accelerator, e.g. from rupture of the electron-permeable window.

2.3 Special Problems

2.3.1 *Ventilation and Ducting*

Accelerator facilities require ventilation for one or more of the reasons discussed in this subsection. At the very least, regardless of the size or power of the accelerator, there should be adequate ventilation for the personnel working in the area, and to remove whatever heat is dissipated in the air from the operation of the machine. Because accelerator facilities are usually isolated from general air circulation systems, ducts are required to introduce and exhaust air in the shielded room. The design and location of ducts in shielding walls are discussed in Section 4.4.2.

a. *Ozone and other noxious gases* are formed by the interaction between any form of ionizing radiation and the air. The magnitude of the problem depends on the absorbed dose rate of the radiation. For example, ozone production is negligible during x-ray therapy or radiography because little of the electron beam power is converted to x rays. However, certain radiotherapy accelerators can also be used for direct electron-beam therapy, which may result in hazardous ozone production, and consideration *shall* be given to means of minimizing this possible hazard.

Electron-beam generators, especially those used for production-line electron irradiation of materials, produce high concentrations of ozone and other noxious gases in the immediate vicinity of the electron-permeable window of the accelerator and the irradiated material. In addition to the health hazard, there is a strong possibility of chemical damage to the accelerator, to the irradiated material, and to any wiring, conduits, pipes, or apparatus in the vicinity. Since the production of these gases can be confined to a relatively small volume of air, it is preferable to exhaust the irradiated air through ducts, the intakes of which are located adjacent to the electron-permeable window. A hood is often used to keep at a minimum the circulation of ozone-laden air into the room. High-capacity blowers are needed to remove the air as fast as possible, and thereby to reduce the possibility of damage. Appendix I includes guidelines on ozone concentrations and their assessment.

Several oxides of nitrogen are also formed in irradiated air. On combination with water vapor in the air, corrosive HNO_3 can be formed, an acid that attacks most metallic surfaces. For this reason, the circulating air *should* be low in moisture content to reduce the corrosive effects of ionizing radiation. In general, health hazards of the oxides of nitrogen are considerably less than the accompanying hazard from ozone.

b. *Airborne radioactivity* may be produced by x rays from accelerators of electrons above 6 MeV in energy, and by neutron-producing accelerators of all types. Again, the magnitude of the problem depends on the radiation emission rate. Radioactivity produced directly in the air from photonuclear reactions is predominantly ^{15}O ($T_{1/2} = 2$ min), ^{13}N ($T_{1/2} = 10$ min), and in some cases ^{16}N ($T_{1/2} = 7s$). Appendix G contains a table of typical photonuclear reaction thresholds.

The air vented from areas in which airborne radioactivity is suspected *shall* be dispersed into the atmosphere in a manner to meet existing local, state, and/or federal air regulations. In particular, dispersion *shall* be planned to eliminate the possibility of the exhaust air being immediately drawn into neighboring air intakes. Radiation areas in which airborne radioactivity can be present *shall* be adequately ventilated before the entry of personnel, where concentrations are expected to exceed permissible limits. Guidance with respect to such limits can be found in NCRP Report No. 22 (NCRP, 1959) and ICRP Report No. 2 (ICRP, 1959). New information and recommendations are continually being generated on this important subject.

c. *Absolute[2] or thorough filtration of the air* is generally not required in accelerator-facility ventilation systems, but in certain specified types of research experiments there may arise a need for such stringent air-filtration methods. For example, radioactive particulates such as uranium and fission products *shall* be filtered on an

---

[2] An "absolute" filter is a particulate filter that has been tested to be 99.97 percent efficient for the removal of 0.3 micrometer, thermally generated DOP (dioctylphthalate) particles.

absolute basis. Machine shops that are associated with accelerator facilities may have to be used, from time to time, in the machining of radioactive or toxic materials. These materials *shall not* be allowed into the general ventilation or air-dispersal system. Machine shops for radioactive materials are usually located separately from other shops in those accelerator facilities that require them.

2.3.2 *Handling, Storage, and Disposal of Radioactive Materials*

a. *Radioactivity induced* by high particle energies and beam powers can be a problem with respect to accelerator components such as cyclotron dee structures, collimating slits, magnet chambers, or beam dumps. In some cases, access to radioactivity areas may have to be limited until the radioactivity has decayed to safe levels. Appendix G includes: (1) a list of typical radioactive isotopes from (γ,n) reactions; (2) a discussion of certain radioactive isotopes induced in cyclotrons by deuteron bombardment; and (3) radioactive decay curves for a 24-MeV deuteron accelerator, to illustrate the nature of the problem. Reference is made to Barbier (1969) for a comprehensive treatment of induced radioactivity. If induced radioactivity is anticipated in an accelerator, the affected components *should* be designed for quick disconnection, encapsulation, and easy removal from the area, if they are to be repaired or replaced. A shielded and sealable storage area *shall* be provided for housing these radioactive components once they are moved out of the accelerator facility. Procedures for the handling and storage of radioactive materials are included in NCRP Report No. 30, *Safe Handling of Radioactive Materials* (NCRP, 1964a). Small items such as radioactive accelerator targets represent a problem in radiation control. Because they are portable and are not different in appearance from the corresponding nonradioactive items, special care *shall* be taken to set up proper procedures in monitoring, tagging, and storing such items.

b. *Radioactive contamination* may result from radioactivity induced in movable and fluid materials such as dust, water, greases, and oils. These materials may become spread about the facility by the movement of personnel or by air circulation. Good housekeeping, dust filtration, and personnel monitoring can keep radioactive contamination under control. Water that is subjected to high-energy irradiation *should* be in a closed-loop system with adequately shielded filters. Any discharge of the water system to sewers or the environment *shall* be monitored for possible radioactivity before disposal. Water leaks *should* be prevented or contained to minimize contamination of the facility. In extreme cases, special drainage into holding tanks may be required to assure that unacceptably high radioactive spills into public sewage systems do not take place. Radioactive scale in cooling-water systems may pose greater problems than the water itself. Water used in beam-dump cooling may accumulate substantial quantities of tritium, with its attendant health hazards (see paragraph *e* below).

   c. *Radioactive products* are intentionally produced in certain types of accelerator facilities, e.g., for activation analysis or as radiochemicals or radiopharmaceuticals. These facilities *shall* be designed to provide adequate means for handling, processing, and storage (paragraph *a* above and paragraph *d* below). Pneumatic tubes or other material-transfer systems that are used to convey radioactive materials out of the irradiation area *shall* be designed so that they do not compromise the effectiveness of the shielding barrier (see Section 4.4.4). Such systems *shall* be designed for safe removal of components or containers that may become stuck during transfer.

   d. *Disposal of radioactive wastes* is the subject of several publications of the International Atomic Energy Agency (IAEA, 1961; 1965; 1967; 1971) and the NCRP (NCRP, 1951a; 1951b; 1954b; 1964a). The designer of an accelerator facility in which significant radioactive wastes will be produced *should* familiarize himself with these documents and should provide adequate storage areas in the design for storing such radioactive wastes prior to their disposal.

   e. *Tritium and other radioactive materials* are often used as targets in nuclear or neutron research. Tritium is of special interest because of its prolific use in small neutron generators, and because of its gaseous nature, its low-energy beta rays (18.6 keV maximum), its property of oxidizing and interchanging with hydrogen in water and most organic materials, and its tendency to become adsorbed on surfaces. Tritium is used in accelerators, not only as a target material but also (in a few instances) as an accelerated particle. Procedures *shall* be adopted that provide for adequate trapping or containment of the evacuated gases from the accelerator vacuum region in which tritium is present. Mercury and/or oil in both diffusion and backing pumps *shall* be carefully inspected for absorbed or assimilated tritium before disposing of the materials or pumps. Ion pumps *shall* be similarly handled. Beam-tube piping *shall* be similarly inspected when changes in the accelerator system are made. The acceleration of tritium ions imposes additional procedures for the safety of personnel. *All* portions of the accelerator system that have been exposed to tritium *shall* be carefully inspected for tritium contamination. In certain cases, even meticulous wipe tests are not adequate to determine the presence of adsorbed or absorbed tritium. For example, the organic seals of some direct-acceleration tubes can retain tritium without detection by conventional tests. Only after the tube has been elevated in temperature, e.g. for repair or modification, is there a significant evolution of tritium. In general, therefore, ion sources, gas bottles, acceleration tubes, and similar components exposed to tritium or tritium ions *should* be considered for disposal as radioactive waste after their useful life. Additional guidelines on facility design and procedures to minimize tritium contamination are to be found in a forthcoming NCRP Report (NCRP, 1977) and Nellis *et al.* (1967).

2.3.3 *Radiation Damage*

Electrical and electronic devices operating within a radiation area are subject to failure as a result of radiation damage, unless special precautions are taken. Hazard to personnel may arise if the accumulated dose absorbed by safety devices renders them inoperative or undependable through internal failures not detectable by casual inspection. A fire hazard may develop if wire or cable insulating materials deteriorate as a consequence of exposure to large radiation fluences. Failures due to radiation damage may be caused either by radiation-induced mechanical or electrical failure of insulating materials, or by paralysis of sensitive electronic circuits in a radiation field.

a. *Radiation damage to insulating materials* is produced above threshold values of absorbed dose that are approximately independent of the nature of the damaging radiation, i.e. x rays, neutrons, electrons, or ions. Charged particles will, in general, dissipate their energy in a smaller volume than neutrons or x rays because of the relatively short and finite range of the former. Great variability exists in the sensitivity of different insulating materials to radiation damage. Plastic materials are relatively sensitive, but ceramics are usually quite insensitive. Typical radiation-damage thresholds are given in Appendix B-4. When the absorbed dose received by a particular insulator or insulating material exceeds its threshold for damage, it may be assumed that a possible malfunction of the material could develop.

b. *Radiation paralysis of electronic devices,* whether temporary or permanent, can be caused by high absorbed dose rates of radiation, through ionization effects or through atomic displacements produced in the circuit components. Pulsed radiation fields may cause circuit paralysis at lower *average* absorbed dose rates because very high instantaneous ionization may occur.

c. *Recommendations* for minimizing hazards due to radiation damage to insulating materials and electrical and electronic devices include the following:

(1) Consideration *should* be given to the use of radiation-resistant insulating materials, such as ceramics or mineral-oil insulated wire, where exposure to large radiation fluences is to be expected.

(2) Electrical or electronic devices, especially with components that are sensitive to radiation damage, *should* be protected from direct or scattered radiations as much as possible. Fail-safe circuits *should* be utilized in safety devices.

(3) Periodic preventive maintenance and/or replacement *should* be undertaken on safety devices, the frequency of which would depend on spot radiation surveys and the radiation-damage thresholds of the irradiated materials.

3. Sources of Radiation

3.1 General

3.1.1 *Radiations from Accelerators*

Whenever charged-particle beams are brought out of the evacuated region of the accelerator, they constitute a radiation hazard *in themselves*.

Any accelerated beam of particles may produce radiation as a consequence of an interaction between the particle beam and the material it strikes. Radiations of the following kinds may be produced from accelerators:

a. *x radiation* (bremsstrahlung): from the impingement of electrons on matter.
b. *characteristic x radiation:* from the impingement of either electrons or ions on matter.
c. *prompt gamma radiation:* from the impingement of ions or neutrons on matter.
d. *neutron radiation:* from the impingement of either electrons, photons, or ions on matter.
e. *delayed radiations* (e.g. beta and gamma rays): from induced radioactivity.

(Note: x rays and gamma rays, being equivalent in their electromagnetic nature, can often be discussed together.)

These radiations stream radially from the area of beam impingement, and they are usually penetrating in nature. Some radiations have very anisotropic angular distributions, not necessarily peaked in the direction of the accelerated charged-particle beam. Radiations commence and stop with the operation of the accelerator except for delayed radiations from the induced radioactivity.

There may be several sources of radiation throughout an accelerator, depending on its design and operating condition. The existence of these various sources may in some cases be due to a malfunction of the accelerator or of its ancillary apparatus. Of particular importance at higher particle energies and currents is the possible production of significant amounts of induced radioactive material, wherever the beam of particles or radiation may impinge.

These radiations and their sources are discussed in this section, insofar as they affect the basis for designing radiation shielding. Conservative estimates of important radiation characteristics of typical accelerators can be derived from this section and Appendices D (electrons and ions), E (x rays), and F (neutrons). These estimates may supplement or confirm information obtained from the accelerator designer or manufacturer.

3.1.2 *Estimation of Radiation Emission Rate*

To estimate the emission rate of radiation from a specific accelerator, the following parameters are of importance:

a. *Species of accelerated particle:* The particle species (electron, positron, or specific accelerated nuclide) governs the classes of radiation-producing reactions that must be taken into consideration in shielding design.

b. *Target material:* The nuclear composition of the target must be known to determine the specific radiation-producing reactions. If the target consists of more than one nuclide species, the fractional contributions to emission rate must be summed. To estimate the maximum radiation emission rate, the target is assumed to be thick enough to stop all impinging particles. The term "target" includes all materials that can be struck by the particle beam, during or after acceleration, inside the accelerator vacuum region or (in the case of externally produced particles) throughout the radiation room.

c. *Energy of accelerated particles:* The emission rate from a specific reaction is a function of the particle energy. In the case of radiation from endoergic reactions, the impinging particle energy must be greater than the threshold energy for the reaction. The particle energy *should* be assumed to be monoenergetic at the maximum value attainable from the specific accelerator.

d. *Particle-beam current:* Radiation emission rate is generally proportional to the beam current striking the target material. In the case of pulsed beams, the time-average current is the product of the average current during the pulse and the duty cycle.

e. *Angular distribution of radiation:* The angular distribution of the radiation emission rate from a target is measured with respect to the direction of the impinging particles, and it varies according to the specific reaction and the particle energy. Angular distributions for several of the more important radiation-producing reactions are given in Appendices E-2 and F-4. It is to be noted that the emission rate of radiations from either electron or ion impingement on targets tends to peak in the forward direction, except for the important case of neutrons from (γ,n) reactions, where the maximum emission rate is in the sideward direction.

f. *Energy spectrum of radiation:* Information concerning the distribution in energy of the radiation from a target, including its magnitude as a function of angle, is of importance in determining the attenuation characteristics of the radiation. This parameter is discussed in detail in Section 4.2. For some situations, the energy spectrum may not be known. Transmission curves have been provided in Appendices E-7 and F-6, in which the transmission of several spectra is plotted in terms of the accelerated-particle energy and the thick target on which it impinges.

g. *Radiation field map of accelerator:* For any specific accelerator, the profile of radiation emission rate may be different from the angular distribution described in paragraph *e* above, by virtue of attenuation of the radiation in the materials of the accelerator. In some cases, particularly for accelerators in a fixed mounting, this inherent attenuation can provide sufficient basis for designing thinner shielding barrier thicknesses than otherwise would be justified. In other cases, the measured profile may indicate a higher dose-equivalent rate than expected, due in part to beam impingement on collimators, magnet chambers, charge strippers, etc.

3.2 Accelerated Charged Particles

3.2.1 *Electrons*

Electrons are produced for acceleration in a beam by emission from the cathode of an acceleration system, either from a hot filament or from an indirectly heated surface. In a few instances electrons are drawn from the plasma of an ion source, e.g., in certain direct accelerators that can produce either ions or electrons. Very high currents of electrons are drawn from cathode material by field emission, e.g., in flash electron/x-ray machines. Electrons can also be produced by the interaction of other electrons or ions with matter, or by field emission in high electric-field gradients. These phenomena are often of importance within the evacuated acceleration region and beam-transport structure. Secondary electrons usually proceed along different paths than that of the impinging particles, thereby giving rise to other possible sources of x rays.

Electrons have a finite range in matter, approximately proportional to their initial energy and inversely proportional to the density of the absorbing material. The maximum electron range is approximately 0.6 g cm^{-2} times the energy in MeV, in the energy range 2 to 20 MeV. Appendix D-1 includes a curve of maximum electron range (in units of g cm^{-2}) as a function of incident electron energy, for several commonly used absorbing materials. This relationship is, for practical purposes, independent of atomic number.

In many applications, the electron beam is brought out of the accelerator into the atmosphere through a thin, low-energy-loss membrane or "window". Since air has a density of about 1.2×10^{-3} g cm^{-3} at normal room temperatures and pressures, the maximum range in air of such electrons is approximately 5 meters times the energy in MeV.

Electrons are scattered by the air and by all materials on which they impinge and, therefore, they constitute a problem in accelerator facility design. Scattering may occur throughout the entire range of the electron beam. *For shielding calculations, the energy of the scattered electrons shall be assumed to be equal to the energy of the accelerated electrons.*

3.2.2 Light Ions (Hydrogen and Helium)

Included in this category are positive and negative ions of the isotopes of hydrogen and helium. Ions are produced for particle acceleration by means of a gas discharge, usually initiated and maintained by a radio-frequency field. These ions are then extracted from the resulting plasma by an electrostatic lens for acceleration by the electric fields within the accelerator.

The ranges of ions in matter are finite, as in the electron case, but considerably shorter; for example, a 2-MeV proton has about 1/100 the range of a 2-MeV electron, in the same material. As a consequence, ions of most species are not generally brought out into the atmosphere, because even very thin metallic windows between vacuum and atmosphere absorb a significant fraction of the ion energy. At energies above 10 MeV, however, there may be some operational value in bringing ion beams out of the vacuum, in which case, protective measures may be necessary. For example, a 20-MeV proton has a range of about 4 meters in air. Typical ion range-energy relationships are given in Appendix D-3.

Light ions can initiate a broad variety of nuclear reactions, even at low beam energies. Practically all reactions yield at least one gamma ray; many reactions produce neutrons; and most reactions result in a residual radioactive nucleus.

Deuterons (^{2}H) represent a special problem in ion acceleration because of the ^{2}H(d,n)^{3}He reaction, i.e., a neutron-producing reaction by deuterons impinging on deuterium. In any deuteron-acceleration system, there is a strong possibility that deuterium will eventually be adsorbed on the surfaces of beam stoppers, targets, collimators, or other beam-line components. A background of neutrons can thus develop as the deuterium builds up on the surfaces.

Tritons (^{3}H) are occasionally accelerated, to produce certain important nuclear reactions. Because tritium is radioactive, it must be well contained in, and carefully retrieved from, the accelerator vacuum system. Furthermore, adsorption of tritium onto removable beam-line components represents a contamination hazard. This subject is discussed in greater detail in Section 2.3.2.

3.2.3 *Heavy Ions (Lithium and Heavier)*

Generally speaking, the heavy-ion beams that are obtainable from accelerators within the energy scope of this report do not *by themselves* constitute a radiation hazard, provided that they are contained within the vacuum region of the accelerator. Furthermore, radiations that are produced by heavy-ion impingement are relatively low in emission rate as compared with other radiations that can be produced by the same accelerator. Characteristic x rays can be produced by heavy-ion impingement on materials, even at energies of a few keV. The dose-equivalent rates are low but they can be hazardous to personnel directly viewing the fluorescence of an ion beam during accelerator tune-up procedures. Neutrons can be produced by heavy-ion bombardment at energies above a few MeV, particularly on hydrogenous targets.

3.3 Penetrating Radiations

3.3.1 *X Rays*

For the purpose of this report, the term "x rays" generally connotes the continuous-spectrum bremsstrahlung produced by the energy lost by electrons as they are decelerated by nuclei. Characteristic x rays, produced either by electrons or by ions as a consequence of the removal of orbital electrons from an atom, are relatively low in energy (less than 100 keV), and their contributions are usually not dominant in radiation-shielding considerations for particle accelerators.

X rays are produced by the impingement of electrons on matter, with emission rates that increase with the atomic number of the target material and even more strongly with electron energy. Below 1 MeV, the emission rate is greatest in the sideward (90°) direction. With increasing electron energy, however, the angular distribution in x-ray emission rate peaks in the forward direction. An important parameter for accelerator shielding design is, therefore, the 90° emission rate (<1 MeV) or the forward (0°) emission rate ($\geqslant$1 MeV) from high-atomic-number (high-Z) thick targets. Appendix E-1 includes a compilation of experimental forward-directed emission rates, together with several 90° values, for electron energies between 0.1 and 100 MeV. The estimation of emission rates for lower-Z materials and for all angles can be determined, for purposes of shielding calculations, from procedures outlined in Appendices E-1, E-2, and E-3.[3]

The (γ,n) nuclear reactions[4] are of particular importance in shielding designs for electron accelerators and x-ray generators that produce radiations greater than 10 MeV in energy. The neutrons from these reactions are discussed in Section 3.3.2. At electron energies above 30 MeV, other more complex nuclear reactions are possible, but the resulting hazards, e.g., from induced radioactivity, are generally

---

[3] *Example:*
Estimate the x-ray emission rate from a 1-cm diameter, 2-mA, 3-MeV electron beam incident on a thick steel (Fe) target, at 0° and 90° to the electron beam direction. Refer to Appendix E-1, and first assume the high-Z target to be tungsten (W):

$$D_0(0°, W) = 1.1 \times 10^3 \text{ rads m}^2 \text{ mA}^{-1} \text{ min}^{-1},$$
$$= 2.2 \times 10^3 \text{ rads m}^2 \text{ min}^{-1}.$$
$$D_0(90°, W) = 3.0 \times 10^2 \text{ rads m}^2 \text{ mA}^{-1} \text{ min}^{-1},$$
$$= 6.0 \times 10^2 \text{ rads m}^2 \text{ min}^{-1}.$$

Refer to Appendix E-3:

$$D_0(0°, Fe)/D_0(0°, W) = 0.7,$$
$$D_0(0°, Fe) = 0.7 \times 2.2 \times 10^3 = 1.5 \times 10^3 \text{ rads m}^2 \text{ min}^{-1}.$$
$$D_0(90°, Fe)/D_0(90°, W) = 0.5,$$
$$D_0(90°, Fe) = 0.5 \times 6.0 \times 10^2 = 3.0 \times 10^2 \text{ rads m}^2 \text{ min}^{-1}.$$

[4] Strictly speaking, these reactions should be termed (x,n) when they are initiated by x rays from accelerators, as is the usual case in this report. However, common usage in the literature is the term (γ,n), whether the reaction is initiated by gamma rays or x rays.

not as severe as those from the (γ,n) reactions, within the particle-energy scope of this report.

Electron or x-ray energies *below* 1.67 MeV (the photodisintegration threshold for beryllium) are insufficient to produce nuclear reactions of importance for radiation-protection purposes. No induced radioactivity or radioactive contamination is associated with these low-energy electron accelerators. The emerging electrons and the x rays constitute the only radiation hazards.

3.3.2 *Neutrons*

Neutrons are produced from many kinds of nuclear reactions, and the emission rates, energies, and angular distributions of the neutrons depend strongly on the species and energy of impinging particle, as well as on the target material. It is therefore necessary to understand the characteristics of each relevant neutron-producing reaction in order to evaluate the magnitude of the associated radiation-shielding problem.

There are several nuclear reactions that cause the emission of monoenergetic neutrons, where the neutron energy depends on the incident-particle energy and the angle of neutron emission. A few reactions produce neutrons in specific groups of energies. Photoneutron reactions yield a continuous spectrum of neutrons, ranging from very low energies to a maximum that is approximately the difference between the x-ray energy and the threshold energy for the specific reaction.

Several of the more prolific neutron-producing reactions are described in Appendices F-1, F-2, and F-3 with respect to their threshold energies, neutron emission rates and energies from thick-target reactions, as a function of incident particle energy.[5]

[5] *Example:*
Estimate the neutron emission rate at 1 meter from a 1-cm diameter, 10-μA, 10-MeV deuteron beam incident on a thick beryllium (Be) target, at 0° and 90° to the deuteron-beam direction. Refer to Appendix F-2 and the ^{9}Be(d,n)^{10}B yield curve:

Total (4π) neutron yield, $Y = 1.6 \times 10^{10}$ s^{-1} μA^{-1}.

Refer to Appendix F-4, and the angular distributions for *all* (d,n) reactions above 5 MeV:

Y sr^{-1}(0°)/$Y \geq 0.5$ (assume = 1 to be conservative),
$\simeq 1.0 \times 1.6 \times 10^{10}$ s^{-1} sr^{-1} μA^{-1},
$= 1.6 \times 10^{11}$ s^{-1} sr^{-1},
$\phi_0(0°) = 1.6 \times 10^7$ m^2 cm^{-2} s^{-1}.

Y sr^{-1}(90°)/Y sr^{-1}(0°) ≤ 0.1, $\phi_0(90°) \simeq 1.6 \times 10^6$ m^2 cm^{-2} s^{-1}.

The (d,n) reactions are, in general, the most prolific neutron producers (see Appendix F-2). The ^{3}H(d,n)^{4}He reaction, for example, provides a high fluence rate of 14-MeV neutrons at deuteron energies above 0.1 MeV and is utilized in most small neutron generators. The ^{9}Be(d,n)^{10}B reaction is a most prolific source of neutrons at deuteron energies above 1 MeV. The ^{12}C(d,n)^{13}N reaction can produce a significant background of neutrons, by interaction between impinging deuterons and carbon-contaminated surfaces in the vacuum region of the accelerator. The ^{2}H(d,n)^{3}He reaction is a source of monoenergetic neutrons in the 2- to 10-MeV range. It also can create a neutron background, due to adsorption of deuterium gas on inner surfaces of vacuum systems. Appendix F-5 shows the trend of neutron energies as a function of incident particle energy for selected reactions.

At low bombarding energies, the (p,n) reactions are not as prolific in the production of neutrons as the (d,n) reactions, by at least an order of magnitude (see Appendix F-1). As the proton energy is increased, however, the neutron yields increase rapidly to magnitudes comparable with those from many (d,n) reactions.

The neutron yields from (γ,n) reactions (Appendix F-3) are also lower in magnitude than those from (d,n) reactions. However, very high fluence rates of neutrons can be produced by high-power electron beams, with neutron energies that approach the impinging electron-beam energy. In (γ,n) reactions, electron-beam power is converted as efficiently as possible to x-ray power (in high-Z targets), which, in turn, is used to irradiate a mass of beryllium (for electron energies below about 10 MeV), or uranium, thorium, or plutonium (at electron energies above 10 MeV). Tantalum and tungsten are commonly used to avoid problems with fission products. The photon interaction with beryllium (or deuterium) is a photo-disintegration reaction; with uranium and similar fissionable materials, photofission reactions also contribute to the photoneutron yield. In the latter case, the x-ray converter is often the fissionable material itself because of its very high atomic number.

Neutrons can be produced from very high-Z targets of electron accelerators used for x-ray radiotherapy and industrial radiography, when the electron energy is above about 8 MeV. Appendix F-3 includes a curve of neutron yield as a function of incident electron energy for the ^{181}Ta (γ,n) reaction (see also Axton and Bardell, 1972).

Radioactivity induced by neutron interactions with matter (whether in the materials of the accelerator, ancillary apparatus, shielding walls, cooling water, or the atmosphere) is frequently a hazard at neutron-producing accelerators (see, for example, Barbier, 1969).

3.4 Other Sources of Radiation

3.4.1 *General*

In addition to the radiations from the intended target or the beam stop of an accelerator, there are other possible sources of radiation throughout the accelerator that must be considered in any facility design, as well as later in operational programs.

Electrons that are reflected from x-ray targets or electron-permeable windows possess sufficient energy to stream back into the acceleration tube. Wherever they impinge on the materials of the tube system, x rays can be produced. In many accelerator designs, the solid angle through which these scattered electrons can stream is held at a minimum, to reduce the possibility of radiation damage to the equipment.

The x rays produced from electron impingement on x-ray targets radiate in all directions. The backward-directed component can be reflected by the material within the acceleration tube and its environment, thus producing additional radiation hazards. The component of direct radiation that proceeds at about 180° to the direction of the electron beam is attenuated only slightly within the accelerator because the acceleration path is free from interposing material. Many accelerators that are used for x-ray generation have, therefore, a radiation field map that is characterized by a high exposure-rate "spike" in the 180° direction.

Materials that are struck by x rays, e.g. collimating diaphragms used with x-ray generators for radiotherapy, may emit sufficient numbers of secondary electrons to modify significantly the radiation exposure of the patient. Steps *should* be taken to minimize this hazard either by facing the fixed diaphragm with low-Z material, or by maintaining a target-to-patient distance great enough to cause absorption and scatter of these low-energy electrons in the intervening air space.

3.4.2 *Direct Accelerators*

In either single-stage or tandem direct ion accelerators, electrons can be accelerated toward the positive high-voltage terminal or midterminal. These electrons are released by ion impingement from the residual gas within the acceleration tube or from the tube-electrode structure. As a consequence, the positive terminal (and sometimes the acceleration-tube electrodes) becomes a source of x rays. Modern acceleration tubes are designed to reduce this flow of secondary electrons, in both energy and current, but they do not completely eliminate it. The magnitude of the electron current varies with the condition of the acceleration tube and its vacuum. A conservative estimate of x-ray emission rate can be derived by assuming that the electron current is of the same magnitude as the ion current (for single-stage accelerators) or twice the magnitude (for tandems); that the electron energy corresponds to the value attained by acceleration by the total potential impressed across the tube; and that the target material for the electrons is steel.

Electrons can occasionally be emitted back into the acceleration tube of positive-ion accelerators, from sharp edges or points on slits, apertures, or lenses in the beam-transport system. These electrons can produce x rays within the acceleration region, as described above. Whenever a change is made in the design or equipment of beam-transport systems, a radiation survey *should* be made to redetermine radiation levels.

The mid-terminal of a tandem-type direct ion accelerator can also be a source of neutrons. The midpoint of the double acceleration process in a tandem is at positive voltage with respect to ground. A charge-changing gas or foil is situated in the mid-terminal to change the accelerated ion from a negative to a positive charge state in flight. A small percentage of the negative ions accelerated to the mid-terminal can impinge on the structural materials of the charge-changing system. With sufficient ion energy (and especially for accelerated deuterons) neutrons can be produced from nuclear reactions between these ions and the materials of the charge-changing system. A conservative estimate of neutron emission rate can be derived by assuming that 10 percent of an accelerated negative deuterium-ion beam is incident on a thick aluminum target within the mid-terminal (see caption of Appendix F-2 for neutron yield estimate from the ^{27}Al(d,n) reaction).

3.4.3 *Spurious Radiations from Accelerators*

Under poor vacuum conditions or while accelerator-vacuum components are being outgassed, there is a distinct possibility that small "dark currents" of electrons will be inadvertently accelerated when the machine voltage or radio-frequency generating system is turned on—even if the ion or electron source is not turned on. It is unsafe, therefore, to depend only on the particle-source control to eliminate the radiation. Accelerators of the direct and electron linear types are particularly prone to "dark current" production during warm-up or conditioning of accelerator components.

Direct accelerators have a capacitance. Until the charge on it has been brought to zero, radiation may be emitted.

The Van de Graaff type of direct accelerator, whether operating with positive or negative potential, can accelerate a beam even if the belt-charging system is turned off. The charging belt, being of an insulating medium, can acquire electric charge frictionally while it is running. The polarity of the charge may vary, depending on the belt material and the surfaces along which it runs. As a consequence, it is hazardous to assume that there is no radiation present, unless the charging belt is stopped.

With microwave electron linear accelerators of the standing-wave design, there is a possibility that electrons back-scattered from the x-ray target can be back-accelerated by the same standing-wave fields that are used for producing the forward electron beam. In presently available designs, the solid angle of back-scattered electrons that can be accepted by the waveguide is quite small. Furthermore, the cathode material, against which the backward-accelerated electrons can impinge, is usually of low-Z material, with correspondingly low efficiency of x-ray production. However, measurements in the 180° direction from the forward beam *should* be made to insure that no significant hazard exists.

Cyclotrons can accelerate ions effectively during only a fraction of the cycle of the radio-frequency accelerating voltage. Ions that are injected during the remainder of the cycle eventually become lost in the dee structure. These ions can induce radioactivity in the dee structure, with a gradual buildup that may ultimately necessitate removal of the dees for a "cooling-off" period.

3.4.4 *Ion Sources and Injectors*

In certain types of cyclotrons, electron linacs, and tandem accelerators, the particle beam accepted by the main accelerator system is formed and pre-accelerated in a separate apparatus. Such an injector is essentially a small version of a direct accelerator, often operating at several hundred kilovolts potential, and it produces its own radiations. In some cases the injector is tuned locally before the entire accelerator is put into operation. Personnel *shall*, therefore, be protected from these low-energy radiations.

Some accelerator systems consist of two or more major acceleration components, such as a tandem accelerator with a cyclotron injector, or vice versa. In these cases, the multiple sources of radiation must be considered jointly in the overall facility design.

Radio-frequency ion sources are often used in small direct accelerators, as well as in the injector systems of larger accelerators. Even though the x rays from these sources are relatively low in energy, they represent a potential radiation hazard.

3.4.5 *Klystrons*

These microwave power amplifiers are often used to generate the radio-frequency accelerating electric fields in electron linear accelerators. Since klystrons in accelerator service operate with pulsed-beam voltages and peak currents typically in the range of 100 to 250 kV and 100 to 300 amperes, respectively, very high x-ray emission rates can be produced. Typically, a shielding thickness of 1 to 2 inches of lead has been found to be adequate in protecting operating and maintenance staff from these emission rates.

Because of the irregular geometry of the klystron, particularly in the region of the rf output waveguide and the beam-collector cooling connections, special care must be taken to avoid radiation leakage due to inadequate overlap or fitting of the shielding material. It is strongly recommended that a wrap-around radiographic film technique be adopted as a radiation survey procedure to detect the presence and position of such emissions.

3.4.6 *Leakage Radiation from Radiotherapy Accelerators*

Substantial radiation shielding in the form of a therapeutic-type protective tube housing is required in radiotherapy accelerators, to limit the dose-equivalent rate from x rays produced at the target, in all directions except that of the useful forward-directed beam. The purpose of this shield is to minimize the irradiation of the patient other than at the point of therapy. For x rays above 8 MeV, the shielding design for x rays may be inadequate to limit correspondingly the dose-equivalent rate from photoneutrons that are also produced in the x-ray target (see Axton and Bardell, 1972). The subject of limiting the dose-equivalent rate of leakage radiation to therapy patients is beyond the scope of this report, but it is currently under consideration by other organizations, including the International Commission on Radiological Protection and the Bureau of Radiological Health of the U.S. Food and Drug Administration.

4. Radiation Shielding

4.1 Shielding Materials

4.1.1 *General*

Any material may be used for shielding against radiations from accelerators, if the thicknesses employed are sufficient to absorb or attenuate the radiations to the required levels. Ordinary concrete is most often used for accelerator-facility shielding, but other materials may prove to be more advantageous under certain circumstances. Appendix H includes a list of commonly used, commercially available materials for shielding purposes.

In selecting a shielding material, the following factors should be evaluated:
  (a) required thickness and weight of material;
  (b) possibility of multiple use, e.g. material that serves both shielding and structural purposes;
  (c) possibility of use as shielding against both neutrons and x rays or gamma rays; thicknesses might differ considerably for neutrons and x rays separately;
  (d) uniformity, consistency, homogeneity of shielding;
  (e) permanence of shielding, e.g. stability against cracking, flaking, sagging, changing composition (see notes below regarding water content of concrete);
  (f) optical transparency, for windows; resistance to radiation darkening, chemical or biological contamination, internal optical scattering;
  (g) cost of material, including installation and maintenance;
  (h) architectural appearance; surface characteristics, ease of cleaning or painting;
  (i) possibility of inducing radioactivity or prompt radiation in the material from continued exposure to radiations.

4.1.2 *Commonly Used Materials*

  a. *Electrons*. Low-Z materials such as aluminum or ordinary concrete are to be preferred, to minimize the production of x rays. Air is also an absorber of electrons, and is often utilized as such in those facilities where the absorption distance can be accommodated or where the production of ozone and other noxious gases does not represent a hazard.

b. *Ions.* High-Z materials such as tantalum or platinum are to be preferred as ion-beam stoppers, to minimize neutron production at ion energies below about 5 MeV. Above this energy, light ions produce neutrons by impingement on most materials.

c. *X Rays and Gamma Rays.* Ordinary concrete is the most widely used material for x-ray structural shielding purposes because of its relatively low cost and its reasonable stability. There are occasional advantages to the use of heavy aggregates such as iron ore, to reduce shielding thicknesses in restricted areas, but the relatively high cost of these special aggregates tends to limit their use for shielding purposes.

Attention must be paid to the actual density of concrete, as compared with the design value. It is difficult, for example, to pour ordinary concrete with a set density of 2.35 g cm^{-3} (147 lb ft^{-3}) without special tamping or packing. Furthermore, concrete tends to dry out to a somewhat lower equilibrium density with the passage of time. This effect is more pronounced with cast concrete blocks than with poured slabs.

Steel is occasionally used as a structural x-ray shielding material, e.g. in the form of plates (for rolling doors), or weldments or castings (for self-shielding electron-beam generators). In those cases where there is a requirement for a localized increase of shielding thickness (e.g. in front of ducts), steel plate can be mounted on a bracket to meet the requirement economically and compactly.

Lead is used for a variety of shielding purposes because the required thicknesses are much less than those for low- or even moderate-Z materials. The thickness advantage for lead is even more pronounced at x-ray energies below about 1 MeV, where the photoelectric absorption process (proportional to Z^5) dominates. At x-ray energies above a few MeV, a similar advantage pertains because of the pair-production absorption process (proportional to Z^2). Within a range of x-ray energies between about 1 MeV and a few MeV, the *weight* of lead shielding does not differ much from that of low- or moderate-Z materials for comparable shielding. However, above and below this narrow range, lead is better even on a weight basis because the absorption processes are proportional to higher powers of Z than is the density.

Lead sheets or plates tend to flow under their own weight, unless supported against a rigid backing, or laminated with structural materials such as wood or steel. Lead blocks or castings are useful for collimating shields. Other high-Z materials, such as tungsten alloys or depleted uranium, are also used for collimation, especially where compact designs are required.

Combinations of shielding materials are often utilized for economy or in those cases where accelerator facilities are to be designed into existing structures. Earth is a useful shielding material against x rays, for example, and it can be mounded up against a concrete retaining wall to provide sufficient shielding thickness. Alternatively, earth can be used as a filler between two concrete retaining walls in those cases where the total shielding requirement would otherwise involve concrete thicknesses that are comparatively more expensive.

In shielding calculations, contributions from existing brick, plaster, or composition walls can be included, provided that any departures from homogeneity of these walls are known. Concrete or tile blocks are often hollow, to reduce their weight while retaining their structural advantages. These inhomogeneities must be taken into account, and the thickness of the *thinnest* sections *should* be used in conservative shielding calculations.[6] Since these building materials are, for the most part, low-Z in composition, it is possible to derive their equivalent concrete thickness by multiplying their actual thickness by the ratio of the densities ($\rho_{material}/\rho_{concrete}$).

d. *Neutrons*. Ordinary concrete is the most frequently used neutron-shielding material because of its hydrogen content and its normally low-Z aggregates. One disadvantage of concrete is its decreasing water content over a period of time. The effect of water content on the neutron-attenuation characteristics of concrete is discussed in Appendix H-3.

Materials with very high absorption cross-sections for neutrons near thermal energies, such as boron or cadmium, are sometimes chosen despite their relatively high cost, to reduce shielding thicknesses. Composite materials, such as borated concrete or boron-aluminum alloys, are useful if they can be incorporated into shielding barriers without compromising the required structural stability or strength.

Wood has a favorable hydrogen content, and it has structural

---

[6] The transmission of a hollow wall is sometimes less than that calculated by using only the thicknesses from the holes to the outside. This decrease in transmission is caused by radiation reduction from the webs or septa between the holes. Some data on x-ray transmission through such walls are described in NCRP Report No. 35 (NCRP, 1970). The shielding effectiveness of ribbed slabs has also been studied (see Green *et al.*, 1972 and Chilton and Morris, 1972).

value. It is occasionally used in temporary shields or as plugs and fillers in neutron-shielding barriers. Wood is subject to dimensional changes, occasioned by temperature or humidity changes in the environment. Compressed or reconstituted woods, such as "Masonite", are not as subject to dimensional changes. In either type of material, however, their resistance to radiation damage is quite poor. Consequently, the use of wood for permanent shielding is not recommended.

Water, paraffin, plastics, and organic oils are hydrogenous materials that are occasionally used for neutron shielding, but they lack structural stability unless supported by rigid containers. Paraffin and organic oils *should not* be used for permanent shielding because of their flammability.

For very thick shielding barriers, the required mass per unit area is greatly dependent on the attenuation characteristics of the incident-neutron spectra. For neutrons with energies much greater than those in a fission spectrum, several practical materials of low- or moderate-Z composition could be considered for the bulk of the shield, whereas the last tenth-value layer could usefully be hydrogenous, to attenuate the remaining low-energy neutron components. Gamma rays are produced in shielding materials by the scattering and absorption of neutrons. The gamma-ray component is particularly significant in very thick shields and for neutrons with energies less than those from fission spectra. The neutron calculational data (Section 4.3.3 and Appendix F) take into consideration the production of gamma rays within the shield.

4.1.3 *Special-Purpose Materials*

There are certain relatively costly materials whose specialized radiation-attenuation characteristics can be advantageously applied to such purposes as filling apertures in shielding barriers, observation windows, and linings to shielding barriers.

Holes in shielding walls, e.g. for the introduction of piping or ductwork, can be filled by packing them with steel or lead wool or small lead pellets (for x rays); with borax, borated paraffin, or borated plasters (for neutrons).

Mortars or plasters containing nuclides with high attenuation characteristics are now commercially available. Such materials have different compositions depending upon whether they are to be used for x-ray or neutron attenuation. They are especially valuable for local cast shielding around very high fluence-rate sources of radiation. In the same general category of materials, lead-loaded putties, ceramics, and diatomaceous earths can be utilized.

Linings to shield walls can sometimes reduce overall shielding requirements. For example, a low-Z lining to a concrete wall can be used to reduce the production of x rays by electron impingement on the wall. Similarly, gypsum linings are occasionally used to reduce thermal-neutron flux densities, particularly in mazes or on outside surfaces of shielding walls in which the neutrons have been attenuated.

Laminates of materials have been studied to some extent and even used occasionally, with the purpose of reducing shielding thicknesses by taking advantage of the differences in build-up and/or absorption of radiations in selected materials. In general, these laminates provide some economy of space but at a relatively high cost. Each case must be considered on its own merits. A qualified radiation-shielding expert *should* be consulted to evaluate the use of laminates.

The natural radioactive background of materials is of concern mainly to designers of accelerator facilities for precise nuclear research, in which radiation background is an ever-present problem that is only aggravated by masses of material containing naturally-occurring radioactivity. Aside from this particular application, there is no necessity to take special care in selecting shielding materials for low content of radioactivity.

Observation windows are generally not used in the types of accelerator facilities that are considered in this report, except for very low-energy x-ray installations. Closed-circuit television systems or optical periscopes in labyrinths are often employed. Occasionally, a combination of mirror with window (behind a shielding partition) is utilized. Windows are usually mounted in concrete walls, and the thicknesses for equivalent attenuation, for the two materials, can differ substantially. The design of the window must, therefore, take this difference into consideration, in tapering or compensating the concrete thickness appropriately. A discussion of transparent materials for observation windows is given in NCRP Report No. 49 (NCRP, 1976). Several types of observation windows are commercially available.

4.2 Parameters of Shielding Calculations

4.2.1 *Maximum Permissible Dose Equivalent (H_M) and Dose Limits*

For the purpose of designing accelerator facilities, the H_M and dose-limit values recommended in Appendix B-1 are interpreted in the following manner:

An average weekly H_M and dose-limit value is derived from presently recommended annual H_M and dose-limit values with the assumption that the dose is accumulated uniformly throughout the period. Thus, for design purposes, the following values may be employed:

    Controlled areas   100 mrem in a week

    Noncontrolled areas   10 mrem in a week

In the absence of dependable scheduling information, the accelerator *shall* be assumed to be in continuous operation at maximum radiation emission rate throughout an 8-hour day for 5 days a week, a total of 40 hours per week. This apparently stringent assumption is justified on the basis that the barrier thicknesses might be excessive by perhaps one or two half-value layers, and that the additional cost of the facility would be relatively small. Furthermore, this assumption is aligned to the desire to keep radiation exposures as low as practicable. If it is planned that the accelerator in question will be in operation in excess of this period, e.g. on a multiple-shift basis, or even on an overtime basis, the schedules of the associated workers *should* be arranged so that no worker occupies the controlled areas surrounding the accelerator facility for more than 40 hours per week. As an alternative, the shielding design *should* be increased so that radiation workers who occupy the controlled areas more than 40 hours per week *shall not* receive more than the H_M. In any event, the shielding for noncontrolled areas *shall* provide adequate protection for the general public, in accordance with the dose limits for noncontrolled areas.

It is usually convenient, therefore, to derive an average *hourly* H_M and dose-limit rate ($\dot{H}_M$) for use as reference in shielding calculations.

    Controlled areas   2.5 mrem in an hour.

    Noncontrolled areas   0.25 mrem in an hour.

4.2.2 *Absorbed Dose Index and Dose-Equivalent Index*

To meet the need for the characterization of ambient radiation levels at any location for purposes of radiation protection, the quantities *absorbed dose index*, D_I, and *dose-equivalent index*, H_I, have been defined in ICRU Report 19 (ICRU, 1971a). In this approach, the maximum absorbed dose in the human body is approximated by the maximum absorbed dose in a sphere of tissue-equivalent material. The approach is useful in circumstances where it is difficult to estimate the maximum absorbed dose and dose equivalent in the externally irradiated human body from measurements in air.

The *absorbed dose index*, D_I, at a point is the maximum absorbed dose within a 30-cm diameter sphere centered at this point and consisting of material equivalent to soft tissue with a density of 1 g cm^{-3}.

The *dose-equivalent index*, H_I, at a point is the maximum dose equivalent within a 30-cm diameter sphere centered at this point and consisting of material equivalent to soft tissue with a density of 1 g cm^{-3}. In general, the maximum values of the quality factor, Q, and the absorbed dose, D, occur at different locations in the sphere. However, $Q_{max}D_I$ is a conservative estimate for H_I. (See also Section (1.3.4.)

4.2.3 *General Equation for Calculating Dose-Equivalent Index Rate*

The fundamental purpose of radiation shielding is to reduce the dose-equivalent index rates from all sources of radiation that converge on a particular reference point, so that the dose-equivalent index rate at the reference point does not exceed the applicable H_M or dose-limit value. Mathematically, this can be stated as follows:

$$\dot{H}_{Id} \leq \dot{H}_M$$

where:

$\dot{H}_{Id}$ is the sum of all dose-equivalent index rates at the reference point,

$\dot{H}_M$ is the applicable H_M or dose-limit rate,

and

$$\dot{H}_{Id} = \sum_j \frac{\dot{F}_{0j} B_j T}{K_j d_j^2}, \tag{1}$$

where

$\dot{F}_{0j}$ is the maximum absorbed dose rate or particle fluence rate from the jth source (whether primary or reflected radiation) at a standard distance of 1 meter; F_0 is expressed as $\dot{D}_0$ (for x rays, rads m² min⁻¹), ϕ_0 (for neutrons, m² cm⁻² s⁻¹),

B_j is the shielding transmission ratio for the radiation from the jth source,

T is the occupancy factor of the area represented by the reference point (see Appendix B-6),

K_j is a dimension-converting constant pertaining to radiation from the jth source.

d_j is the distance between the jth radiation source and the reference point (meters).

Specific equations of the type generalized by Equation (1) are given in Section 4.3 for the cases of x rays and neutrons, both primary and reflected.

To calculate the thickness and configuration of shielding barriers required to attenuate radiations to levels below the requisite H_M or dose limit, the values of all of the parameters described below in this section must be numerically established. Methods of calculating shielding thicknesses are given in Section 4.3, in terms of these parameters and with reference to pertinent graphical and tabular material in the appendices.

4.2.4 *Radiation Source Parameters*

a. *The types of radiation* and the *locations of radiation sources* that *shall* be considered in shielding design are diagrammed in Figure 1 for typical electron and ion accelerators. Each type of radiation *shall* be considered separately in shielding calculations, to determine which radiation represents the dominant shielding problem, and/or to permit summing of all dose-equivalent index rates beyond the shield. Each rate is determined by a variation of Equation (1).

b. *The emission rate of radiation* from an accelerator can be estimated from selected data presented in Appendices E-1 to E-6 and F-1 to F-5, as explained in Section 3. These data may tend to overestimate the magnitude of the emission rate, but generally not by more than a factor of two. Where angular distribution data are not available, it is necessary to assume that both the energy spectrum and the emission rate are invariant as a function of angle, and to use the maximum emission rate and known energy spectrum as applying to all angles.

c. *Reflected radiation* from shielding barriers and other materials inside the radiation enclosure is estimated by methods explained in Section 4.3.

4.2.5 Geometrical Factors

a. *The distances between radiation sources and occupiable areas* must be determined, because radiation emission rates from point sources are reduced in magnitude by the inverse square relation. A "point source" is one that has dimensions one or more orders of magnitude smaller than the distance between source and point of observation. Extended sources of radiation can be divided into a number of point sources for separate analysis. The contribution to dose-equivalent index rate at a given reference point from such a source is the sum of contributions calculated for each point source.

b. *The direction of the accelerated particle beam* must be known for purposes of calculating shielding barriers. Except for most photoneutron sources and for x-ray sources from electrons below about 1 MeV in energy, the maximum radiation emission rate is in the same direction.

c. *The floor thickness shall* be determined by the shielding requirements, if there is any type of occupancy below (see Figure 2). If the facility is at grade level or in a basement, however, there is no need to provide special shielding in the floor. There may, of course, be an experimental or application requirement for special materials in the floor, e.g. to reduce neutron scattering or x-ray production from electron impingement.

d. *The ceiling thickness shall* be determined by the shielding requirements, if there is any type of occupancy above the facility (see Figure 2A and 2B). Even if there is no possible occupancy, however, a study *shall* be made to determine whether nearby occupiable buildings would be exposed to radiation through the ceiling, either directly from the source or by skyshine. If so, the ceiling barrier *shall* be sufficient in thickness to attenuate the radiations to the appropriate H_M or dose-limit levels at occupiable locations in these building sites. Alternatively, the barrier walls of the facility *shall* be designed high enough to cast a protective "shadow" over the building in question. In some facility designs that are one-story in construction and remotely located, the ceiling may have been designed only for protection against the weather. In such a case, skyshine, i.e. radiation reflected by the atmosphere, may be a problem *within* the facility. For such situations, the shielding calculations are described in Section 4.5.1.

4.2.6 Workload and Occupancy Factors

There are certain types of accelerator installations, notably radiotherapy facilities, where the production of radiation can be predicta-

4. RADIATION SHIELDING

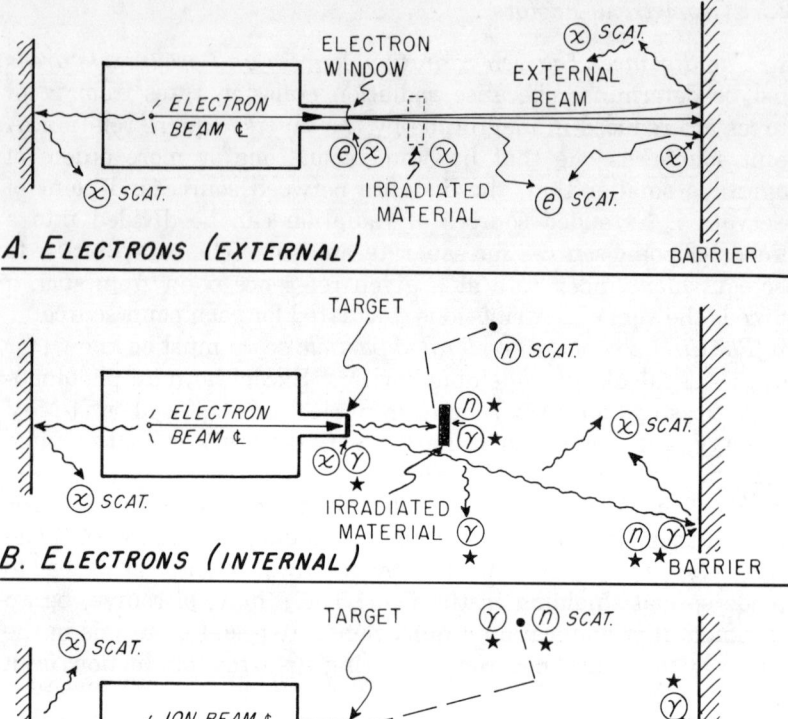

Fig. 1. Radiations from Accelerators. Sources of radiation from several types of accelerators are indicated by the following symbols:

e	electrons
x	x rays
n	neutrons
γ	gamma rays

A star (*) accompanying a symbol indicates that the radiation is produced only at energies above the pertinent nuclear-reaction threshold energy.

₵	centerline of beam

Scattering of gamma rays is not considered important in radiation shielding designs because the other radiations from accelerators dominate the situation.

4.2 PARAMETERS OF SHIELDING CALCULATIONS / 47

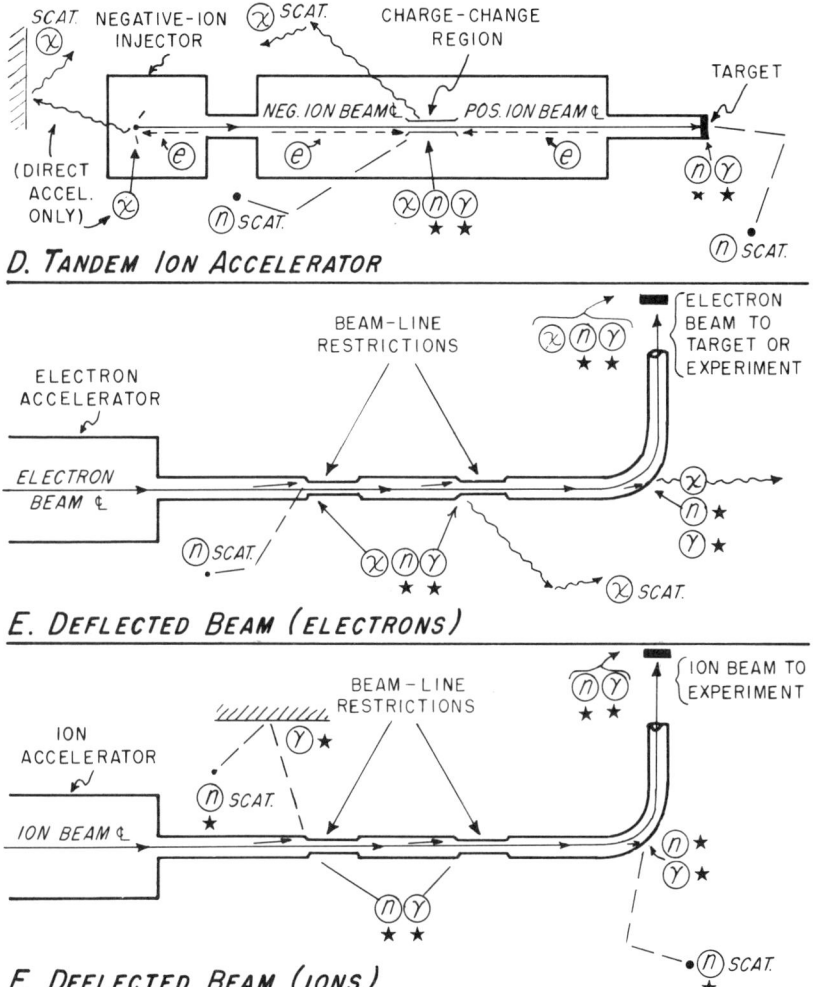

D. TANDEM ION ACCELERATOR

E. DEFLECTED BEAM (ELECTRONS)

F. DEFLECTED BEAM (IONS)

bly restricted to a series of short time periods separated by intervals during which radiation is not produced. NCRP Report No. 49 (NCRP, 1976) utilizes a workload parameter, W, in shielding calculations for radiotherapy x-ray generators. For x-ray generators operating at electron energies of 4 MeV and higher, W is expressed in terms of roentgens per 40-hour week (R/week at a meter or R m^2 week^{-1}). For the purposes of this report, the number of R is equal to the number of rem at the point of interest. Appendix B-5 includes a recommended value for W, for x-ray therapy generators operating at electron ener-

48 / 4. RADIATION SHIELDING

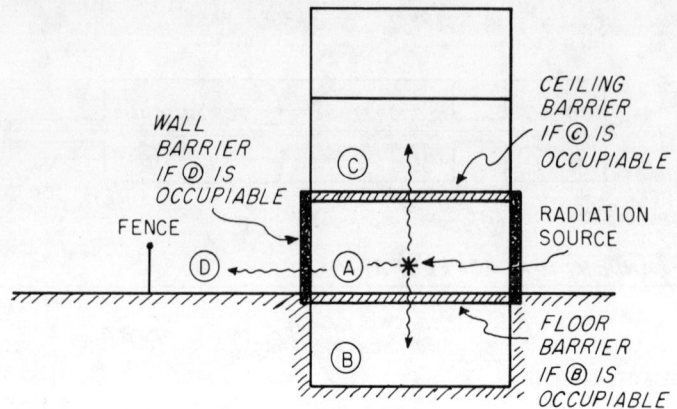

A. Occupancy in Adjoining Areas

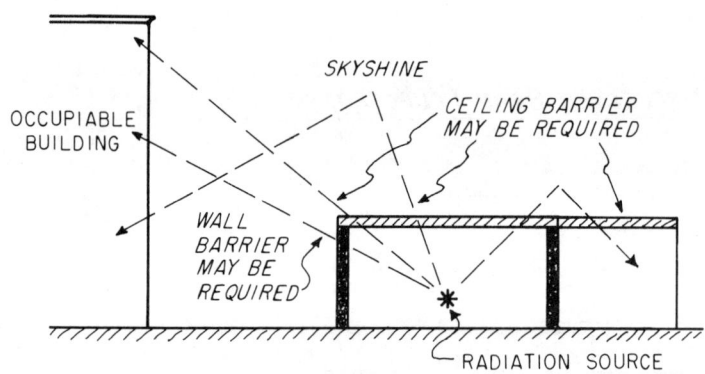

B. Occupancy in Nearby Buildings

Fig. 2. Shielding for occupiable areas.

gies above 10 MeV. Section 4.3.2 includes a special method for calculating shielding-barrier thicknesses for radiotherapy machines, utilizing the parameter W.

Certain classes of noncontrolled areas surrounding an accelerator facility can be assumed to be occupied by the general public for a fraction of the total weekly period during which the radiation under consideration is directed at the area. An occupancy factor, T, can be introduced to modify the maximum radiation emission rate used in shielding calculations. Typical values for T are described in Appendix B-6.

4.3 Methods of Calculating Shielding Thickness

4.3.1 *Calculations for Charged-Particle Beams (Electrons and Ions)*

The required shielding thickness for charged-particle beams is essentially independent of the beam-current density impinging on the absorber. Despite the fact that charged particles lose energy and ultimately become absorbed in air, it is safer to ignore any energy loss in calculating shielding thicknesses, unless the distance between accelerator and the location to be shielded is equivalent to, or greater than, the maximum range of the particles in air.

Electron ranges in air, water, tissue, aluminum, and lead are plotted as a function of incident particle energy in Appendix D-1. To a first approximation, electron ranges are inversely proportional to the density of the absorbing material. Information on the backscattering of electrons is given in Appendix D-2.

Proton ranges are plotted in Appendix D-3, as a function of proton energy, for aluminum, copper, lead, and uranium. Corresponding ranges for deuterons, tritons, ^{3}He, and ^{4}He ions can be calculated from the proton range by means of conversion formulae included in Appendix D-3.

Even though charged particles in an external beam can be completely absorbed in shielding materials, additional thickness may be required in the barrier to attenuate the secondary radiations produced in the shield from the various processes by which the impinging particles become absorbed. X rays, gamma rays, and/or neutrons can be produced by particle interactions with the shielding material. Shielding-barrier calculations for these penetrating radiations are discussed below, in Sections 4.3.2 and 4.3.3. The source of these radiations is, of course, the volume (irradiated area times the particle range) of the shielding material.

4.3.2 *Calculations for X Rays*

X rays are attenuated approximately exponentially through materials. The dose-equivalent index rate is diminished only fractionally through each incremental thickness of material, and it theoretically never becomes zero. It is necessary, therefore, to determine the shielding transmission ratio for x rays, B_x, by which the dose-equivalent index rate *shall* be diminished through the shielding-barrier thickness to appropriate H_M and dose-limit levels. This shielding transmission ratio governs the thickness of the shielding barrier, as discussed later in this Section.

a. *Primary X Rays.* When the primary x rays dominate the shielding situation (see Section 4.2.3), then for a point source of x rays,

$$\dot{H}_{\text{ld},x} \leq \dot{H}_M = \frac{\dot{D}_{\text{Io}} B_x T}{(1.67 \times 10^{-5}) d^2} \qquad (2)$$

$$B_x \geq (1.67 \times 10^{-5}) \left[\frac{\dot{H}_M d^2}{\dot{D}_{\text{Io}} T} \right] \qquad (3)$$

where

B_x is the shielding transmission ratio for x rays; the value by which the x-ray absorbed-dose index rate that is incident on the entrance face of the shielding barrier *shall* be diminished by the barrier thickness to the requisite levels of H_M and dose-limit rate at the exit face of the barrier.

$\dot{D}_{\text{Io}}$ is the absorbed-dose index rate (rads m² min⁻¹) at a standard reference distance of 1 meter from the source.

$\dot{H}_M$ is the maximum permissible dose-equivalent or dose-limit rate (see Section 4.2.1) (mrem h⁻¹).

d is the distance between x-ray source and reference point (meters).

T is the area-occupancy factor (see Appendix B-6).

The constant $(1.67 \times 10^{-5}) = (1 \times 10^{-3})$ rad mrem⁻¹ $\times (1.67 \times 10^{-2})$ h min⁻¹.

For radiotherapy facilities, the following expression may be used:

$$B_{xt} \geq (1 \times 10^{-3}) \left(\frac{\dot{H}_{Mt} d^2}{W T} \right) \qquad (4)$$

where

B_{xt} is the shielding transmission ratio for x rays from radiotherapy sources.

W is the workload (rem m² week⁻¹).

$\dot{H}_{Mt}$ is the H_M or dose-limit rate expressed in mrem week⁻¹, the time interval commonly used in radiotherapy operations (see Section 4.2.6).

Transmission curves of broad-beam x-ray geometries are presented in Appendices E-7 to E-11, in which the shielding transmission ratio, B_x, is given as a function of shielding thickness for a selection of electron energies producing thick-target x rays from very high-Z materials. In all cases, the effects on dose-equivalent index rate due to modification of x-ray spectrum in the shield are implicit. For a calculated B_x, the thickness of shielding barrier can be read directly

4.3 METHODS OF CALCULATING SHIELDING THICKNESS / 51

from these transmission curves, for the materials and electron energies represented. Curves are presented for ordinary concrete, steel, and lead.

For x-ray energies other than those given in the transmission curves, B_x can be related to the shielding thickness in terms of the number of tenth-value layers of the shielding material that are required to diminish the radiation to H_M or dose-limit levels. A tenth-value layer is that thickness through which the x-ray dose equivalent is diminished by a factor of 10. Hence,

$$B_x = 10^{-n}; \text{ or } n = \log_{10}(1/B_x) \quad (5)$$

where n is the number of tenth value layers.

The slope of the semi-logarithmic plot of broad-beam x-ray transmission vs. thickness varies somewhat as the thickness is increased, but the most significant change occurs during the first decade of transmission. Subsequent changes are relatively small, even over several decades of transmission. Consequently, a value for an "equilibrium" tenth-value layer can be conservatively estimated for purposes of calculating shielding-barrier thicknesses, S.

$$S = T_1 + (n - 1) T_e; \quad (6)$$

where

S is the shielding-barrier thickness;

T_1 is the first tenth-value layer in the shielding thickness, facing the radiation source;

T_e is the subsequent tenth-value layer, approximately constant in value, or "in equilibrium".

Values of T_1 and T_e for ordinary concrete, steel, and lead are plotted in Appendices E-12 to E-14 as a function of the energy of the electrons incident on a thick radiation-producing target.[7] Implicit in

---

[7] *Example:*

Calculate the concrete shielding-barrier thicknesses for the forward-directed (0°) and sideward-directed (90°) x rays from a 1-cm diameter, 2 mA, 3-MeV electron beam incident on a thick high-Z (e.g. tungsten) target. The distances between target and reference points outside the barriers are 5 meters (0°) and 3 meters (90°). The space outside each barrier is considered to be a controlled area, with an occupancy factor T = 1. Estimates of x-ray emission rates are obtained from the procedure and example given in Section 3.3.1. Using Equation (3),

$$B_x(0°) = (1.67 \times 10^{-5}) \left[\frac{(2.5)(5^2)}{(2.2 \times 10^3)(1)} \right]$$
$$= 4.7 \times 10^{-7}.$$

This value of B_x is too small to be read from the 3-MeV transmission curve in Appendix E-8. However, the corresponding tenth-value layers ($T_1 = 26$ cm; $T_e = 23$ cm) are obtained from Appendix E-12. From Equation (5),

4. RADIATION SHIELDING

these values are considerations of modification of x-ray spectra in the shielding thickness.

For low-Z shielding materials other than those presented in the curves mentioned above, methods of deriving appropriate barrier thicknesses are described in Section 4.1.

When a shielding barrier is struck by an externally produced electron beam, the electrons generate x rays at or near the surface of that barrier. This spot must be considered as the source for these x rays. The lateral extent of the x-ray source depends directly on the dimensions of the electron beam incident on the barrier. If the source diameter is less than 10 percent of the shielding-barrier thickness, the x-ray source may be treated as a point source. The emission rate can be calculated from knowledge of the electron energy and current, and the atomic composition of the shielding material. Attenuation by the inverse-square relation is calculated by reckoning distances from the shield face, *not* the accelerator (see Figure 3).

If the size of the extended x-ray source is greater than 10 percent of the probable shielding-barrier thickness, the calculational method in Section 4.2.5a may be used to determine the emission rate per unit area from such a source to the reference point. For such special cases, it is recommended that the services of a radiation-shielding expert be enlisted.

$$n = \log_{10} (1/4.7 \times 10^{-1}) = 6.33$$

From Equation (6),

$$S = 26 + 5.33 \times 23 \text{ cm} = 149 \text{ cm or } 59 \text{ in.}$$

A concrete thickness of 167 cm (66 in) is recommended.

A similar procedure is followed to calculate the barrier thickness for the sideward-directed beam.

$$B_x(90°) = (1.67 \times 10^{-5}) \left[\frac{(2.5)(3^2)}{(6.0 \times 10^2)(1)} \right]$$
$$= 6.3 \times 10^{-7}.$$

The transmission characteristics of the sideward-directed x-ray beam differ from those of the forward-directed beam, as shown in Appendix E-6. The tenth-value layers are closely similar to those for x rays in the forward direction from 2-MeV electrons. Referring to Appendix E-12, $T_1 = 22$ cm; $T_e = 20$ cm.

$$n = \log_{10} (1/6.3 \times 10^{-7}) = 6.20.$$
$$S = 22 + 5.20 \times 20 \text{ cm} = 126 \text{ cm or } 50 \text{ in.}$$

A concrete thickness of 137 cm (54 in) is recommended.

Note that an additional thickness is recommended in each of the above calculations, equivalent to at least one half-value layer and chosen to utilize dimensions normally given for constructing barriers of this material.

4.3 METHODS OF CALCULATING SHIELDING THICKNESS / 53

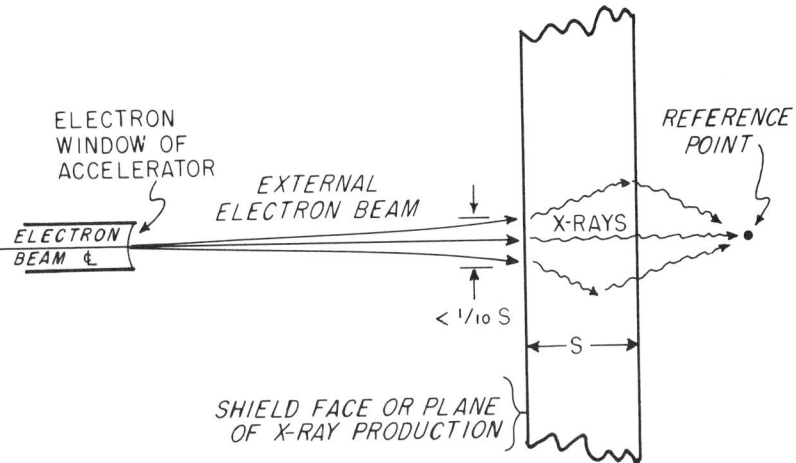

Fig. 3. X-ray production from electrons incident on a shielding barrier.

b. *Reflected X Rays.* Reflected x rays produced within the radiation enclosure must be taken into consideration in the design of mazes and ducts, as well as in the calculation of shielding-barrier thicknesses. Backreflected x rays may be relatively low in energy and absorbed-dose rate, but they can dominate shielding calculations in certain instances where local shielding attenuates the sideward and backward-directed primary x rays, e.g. in radiotherapy and radiography. Referring to Figure 4, three parameters are of importance in calculating shielding-barrier thicknesses for reflected x rays: (1) absorbed-dose index rate of the reflected x rays; (2) distance between reflecting material and reference point; and (3) the transmission characteristics of the reflected x rays through a shielding barrier. The following relationship provides an estimate of the absorbed-dose index rate of reflected x rays as a function of reflecting material:

$$\dot{D}_{I,ro} = \dot{D}_{Io}\, \alpha_x\, A/d_i^2 \qquad (7)$$

where
$\dot{D}_{I,ro}$ is the absorbed-dose index rate of reflected x rays at a distance of one meter from the reflecting area of the material (rads m² min⁻¹);

α_x is the reflection coefficient of reflecting material, dependent on incident x-ray energy, reflecting angle, and reflecting material;

A is the area of reflecting material illuminated by the incident x-ray beam (m²);

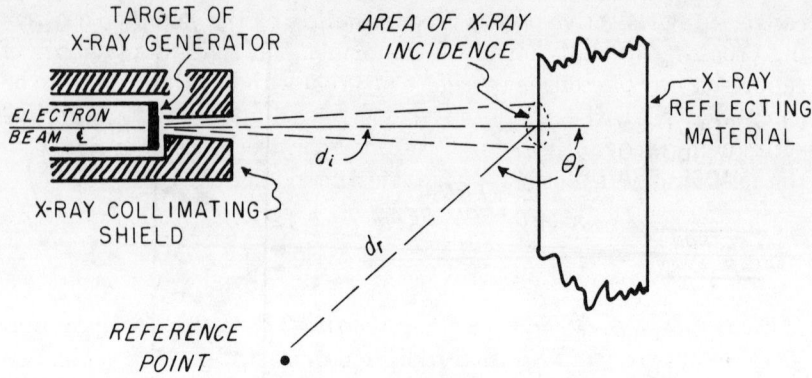

Fig. 4. X-ray reflection geometry for normal incidence.

d_i is the distance from x-ray producing target to reflecting material (meters).

When reflected x rays dominate the shielding situation, the shielding transmission ratio, B_{xr}, is obtainable through a modification of Equations (2) and (3), replacing $\dot{D}_{Io}$ of these equations by $\dot{D}_{I,ro}$ of Equation (7), as follows:

$$\dot{H}_{I,d_r,x} = \frac{\dot{D}_{Io} \, \alpha_x \, A \, B_{xr} \, T}{(1.67 \times 10^{-5}) \, d_i^2 \, d_r^2} \leq \dot{H}_m \tag{8}$$

$$B_{xr} \leq (1.67 \times 10^{-5}) \left[\frac{\dot{H}_M \, d_i^2 \, d_r^2}{\dot{D}_{Io} \, \alpha_x \, A \, T} \right] \tag{9}$$

where d_r is the distance from reflecting material to reference point (meters).

Values of α_x are plotted in Appendix E-15, as a function of *monoenergetic* photon energy (0.1 to 10 MeV), for several reflection angles from ordinary concrete and iron (or steel). An upper limit for α_x for lead is also included in the caption for any energy and angle of reflection. To estimate α_x for a *spectrum* of x rays incident on the reflecting material, it is adequate to assume that the "effective monoenergetic photon energy" of the spectrum is about one-half the electron energy producing the x rays.

To calculate the shielding-barrier thickness required from B_{xr}, it is necessary to select the appropriate transmission curve from Appendix E-7, as follows:

(1) *X rays from electron energies less than 0.5 MeV*: The transmission of the reflected radiation through a barrier is assumed to be the same as that of the x-ray beam incident on the reflecting material.

(2) *X rays from electron energies 0.5 to 3 MeV*: In this energy

4.3 METHODS OF CALCULATING SHIELDING THICKNESS / 55

range, x-ray reflection occurs predominantly by the Compton scattering process, especially for low-Z materials. The maximum energy of reflected x-ray photons is therefore no greater than values that can be calculated from the Compton scattering relation. A conservative estimate of the transmission of reflected x rays can be obtained from the following transmission curves in Appendix E-7:

Reflection Angle	Transmission Curve
90°	0.5 MeV
180°	0.25 MeV

(3) *X rays from electron energies 3 to 10 MeV*: In this energy range, electron-positron annihilation radiation must be taken into consideration. Hence, the 0.5-MeV transmission curves in Appendix E-7 *should* be used for all reflecting angles 90° or greater.

(4) *X rays from electron energies greater than 10 MeV*: For incident x rays from electrons above 10 MeV in energy, the available information concerning the characteristics of reflected x rays is inadequate to provide more than a very conservative recommendation for estimating either α_x or transmission through a shielding barrier. To be on the safe side, therefore, α_x is assumed to be equivalent to the 10-MeV (photon energy) values for ordinary concrete, and 1×10^{-2} in magnitude for iron. The transmission of these x rays is assumed to be the same as the case for the sideward-directed x rays produced by the electron beam.

4.3.3 Calculations for Neutrons

As with x rays, neutrons are attenuated approximately exponentially through thick shielding barriers. Unlike the x-ray case, however, the quality factor, Q, for neutrons varies significantly with neutron energy and transmission through materials. Furthermore, neutrons generate gamma rays within the shield material as they scatter and become absorbed. Consequently, the dose-equivalent index at the exit face of the shielding barrier includes contributions from gamma rays as well as neutrons, to a relative extent that depends primarily on the incident neutron energies, and secondarily on the atomic composition of the shielding material.

a. *Primary Neutrons*. When primary neutrons dominate the shielding situation, the shielding transmission ratio for neutrons, B_n, can be derived from the neutron fluence rate, ϕ_0, as follows:

$$\dot{H}_{\text{Id},n} = \frac{\phi_0 \, B_n \, T}{(2.8 \times 10^{-7}) \, d^2} \leq \dot{H}_M \qquad (10)$$

4. RADIATION SHIELDING

$$B_n \leq (2.8 \times 10^{-7}) \left[\frac{\dot{H}_M d^2}{\phi_0 T} \right] \tag{11}$$

where

B_n is the shielding transmission ratio (rem cm^2)[8];

ϕ_0 is the neutron fluence rate at a standard reference distance of one meter (m^2 cm^{-2} s^{-1});

d is the distance between neutron source and reference point (meters);

The constant $(2.8 \times 10^{-7}) = 2.8 \times 10^{-4}$ h/s $\times 10^{-3}$ rem/mrem.

Multi-collision dose-equivalent index transmission curves for unit fluence of incident monoenergetic neutrons of thermal to 100 MeV energies are given in Appendices F-6 to F-10, derived from recent discrete-ordinate calculations that include contributions to dose-equivalent index by gamma photons produced within the shield (Wyckoff and Chilton, 1973). The shielding thicknesses are expressed in terms of g cm^{-2}, but they are based on the atomic constituency of ordinary concrete (see Appendix H-2)[9]. If the spectrum of incident neutron energies is known, these curves can be used to obtain the contribution to dose-equivalent index from each energy component of the spectrum, which may then be summed to give the total dose-equivalent index from all radiation, implicitly including the contribution from the gamma rays produced by the neutrons in the shield material.

If the incident neutron spectrum is not known, the transmission curves in Appendices F-8 and F-9 may be useful. These represent the dose-equivalent index produced by characteristic neutron spectra from a number of accelerators, as specified in Appendix F-7, in terms of the electrical parameters and nuclear reaction producing the particular spectrum. These transmission curves were obtained by folding the source-neutron spectral distributions into the monoenergetic-neutron data in Appendix F-6.

[8] B_n has dimensions in terms of rem cm^2 because it is the ratio of dose-equivalent index beyond the shield to incident neutron fluence. On the other hand, B_x (Section 4.3.2) is dimensionless, being the ratio of dose-equivalent index beyond the shield to dose-equivalent index without the shield. It is convenient, however, to utilize the symbol B for both ratios because of the similarity in the methods of deriving shielding thicknesses for x rays and neutrons.

[9] The transmission curves in Appendices F-6 to F-10 differ from those given in NCRP Report No. 38 (NCRP, 1971a). The curves in Report No. 38 refer only to tissue kermas due to neutrons, rather than to multi-collision dose-equivalent index due to both neutrons and gamma rays. Figure 60 of that report may be used to evaluate the tissue kerma due to gamma rays, but the sum of the two kermas may still differ from those derivable from the present report because of new information on cross-sections and a different assumed concrete composition.

4.3 METHODS OF CALCULATING SHIELDING THICKNESS / 57

If the calculated neutron transmission ratio is smaller than can be accommodated by the curves in Appendix F-6, F-8, or F-9, appropriate numbers of tenth-value layers can be added to the maximum thickness derivable from the curves. A curve of tenth-value layers as a function of monoenergetic neutron energy is given in Appendix F-10, and tenth-value layers for selected neutron spectra are given in Appendix F-7. All values were empirically derived from the respective transmission curves. *They are applicable only for transmissions less than 10^{-16} rem cm².*

In the transmission curves and tables referred to above, the neutron quality factor has been taken into consideration, and *all* transmission curves in Appendix F are normalized to the dose-equivalent index per unit neutron fluence (rem cm²) incident on the face of the shield. The above transmission calculations assume a broad parallel beam of incident neutrons. This assumption does not seriously affect shielding-barrier calculations in which the incident neutrons are assumed to be emitted from a point source, provided that the distance between source and reference point is greater than the shielding-barrier thickness.[10] *Note: The accuracy of the neutron transmission data may be no better than a factor of 2, and a half-value layer of shielding material* should *be added to the calculated thickness.*

The thickness of the shield may vary from wall to wall around the

---

[10] *Example:*

Calculate the concrete shielding-barrier thicknesses for the forward-directed (0°) and sideward-directed (90°) fast neutrons from a 1-cm diameter, 10-μA, 10-MeV deuteron beam incident on a thick beryllium (Be) target. The distances between target and reference points outside the barriers are 5 meters (0°) and 3 meters (90°). The space outside each barrier is considered to be a controlled area, with an occupancy factor, T = 1. Estimates of neutron emission rates are obtained from the procedure and example given in Section 3.3.2. Using Equation (11),

$$B_n(0°) = (2.8 \times 10^{-7}) \frac{(2.5)\,(5^2)}{(1.6 \times 10^7)\,(1)}$$
$$= 1.1 \times 10^{-12}.$$

Appendix F-9, Curve C (16-MeV) *should* be used to determine the concrete shielding-barrier thickness, since a 10-MeV curve is not available. From Curve C, a thickness of 370 g cm⁻² is obtained. For ordinary concrete of density 2.35, this value represents a linear thickness of 157 cm (62 in). Recommended thickness is 168 cm (66 in).

A similar procedure is followed to calculate the barrier thickness for the sideward-directed beam.

$$B_n(90°) = (2.8 \times 10^{-7}) \frac{(2.5)\,(3^2)}{(1.6 \times 10^6)\,(1)}$$
$$= 3.9 \times 10^{-12}.$$

Using Curve C again, a thickness of 320 g cm⁻² is obtained, corresponding to a linear thickness of 136 cm (54 in) or a recommended thickness of 152 cm (60 in).

facility, not only because of the inverse-square relation, but also because of the angular distribution in fluence rate of neutrons from the target. For incident ion energies below a few MeV, the angular distribution of neutrons tends to be isotropic. However, at higher ion energies, neutron fluence rates tend to peak in the forward direction, in some cases by as much as a factor of 10. For photonuclear processes, the reverse is true, and the fluence rate at 90° may be as much as a factor of 2 greater than at 0°, particularly for thin targets. At very high (above ≈ 100 MeV) impinging electron energies, however, the *highest-energy* neutrons emitted from the target tend to be in the forward direction. Appendix F-4 includes estimates of angular distributions of emission rates for several typical neutron-producing reactions.

b. *Reflected Neutrons.*

(1) *General (Multiple) Reflection within Large Facility.* Reflected neutrons within an accelerator facility may contribute to the total fluence rate at the entrance face of the shielding barrier, and calculations of barrier thicknesses *should* take this possibility into consideration. Reflected neutrons arise mainly from backscattering off the various barriers surrounding the accelerator. The spectra of these neutrons are degraded with respect to the incident spectra. At walls more than a few meters away from the neutron source, the total neutron fluence rate from reflection by other wall surfaces may approach an order of magnitude greater than the fluence rate from primary neutrons at the wall in question. However, because of differences in quality factor for incident and reflected neutrons, the contribution to dose-equivalent rate by reflected neutrons may be only a factor of two. *It is recommended that an additional half-value layer of shielding material be included in the calculated shielding thicknesses for all barriers in the facility, to attenuate these multiply reflected neutrons.*

(2) *Reflection from Collimated Neutron Beam.* In some cases, the reflected neutrons are the result of the backscattering from a *collimated* beam of neutrons incident on a well-defined area of wall that is small in comparison to the total wall area. Should such a situation arise, the neutrons reflected from this area may be the dominant radiation on all walls that are not struck by the primary collimated beam. The source of reflected neutrons, analogous to Equation (7), is described as follows:

$$\phi_{so} = \phi_0 \, \alpha_n \, A/d^2 \qquad (12)$$

where

ϕ_{so} is the fluence rate of reflected neutrons at distance of one meter

from the backscattering source (m^2 cm^{-2} s^{-1});
α_n is the neutron reflection coefficient;
A is the area of the backscattering source (m^2).

Values of α_n are plotted in Appendix F-12, as a function of monoenergetic neutron energy (0.1 to 14 MeV), for several reflection angles from ordinary concrete and iron (or steel). To estimate α_n for a narrow spectrum of neutrons incident on the reflecting material, the median energy in the spectrum can be used. For a broad spectrum, the maximum value of α_n at the angle in question should be used.

In addition to the fluence rates at the shielding barrier from the single reflection, contributions are to be expected from multiple reflections as in (1) above. *An additional half-value layer of shielding material* should *be included in the calculated shielding thicknesses for all barriers to attenuate these multiply reflected neutrons from a collimated neutron beam.*

4.3.4 *Calculations for Multiple Sources*

Certain direct accelerators can be used optionally for the acceleration of either electrons or positive ions, depending on the polarity of the accelerating potential. Referring back to Figure 1, it is apparent that the locations and kinds of radiations differ for the two modes of operation. Shielding-barrier thicknesses *shall* be calculated for all radiations, to determine which source of radiation requires the most barrier thickness. The *maximum* thickness for each barrier *shall* be used in the facility.

With large complex accelerators, in which two or more injectors or accelerator systems are operated in series (see Figure 5), there may exist a multiplicity of radiation sources operating simultaneously. The shielding requirements for all sources *shall* be calculated separately and compared. If it is suspected that the calculated shielding-barrier thickness, S, for each of the j sources will be comparable, a conservative value for S can be obtained by using H_M/j in the individual shielding calculations and selecting the maximum value of S so derived. If, on the other hand, the maximum calculated value of S is thicker by at least a tenth-value layer for the most penetrating radiation involved, this thickness would be considered to be adequate for shielding-design purposes. For intermediate cases, it may be necessary to use Equation (1) and iterate the calculation for the maximum required shielding-barrier thickness.

In any case, S is the barrier thickness as measured along the radiation path through the shielding wall. The wall thickness is $S/\cos$

4. RADIATION SHIELDING

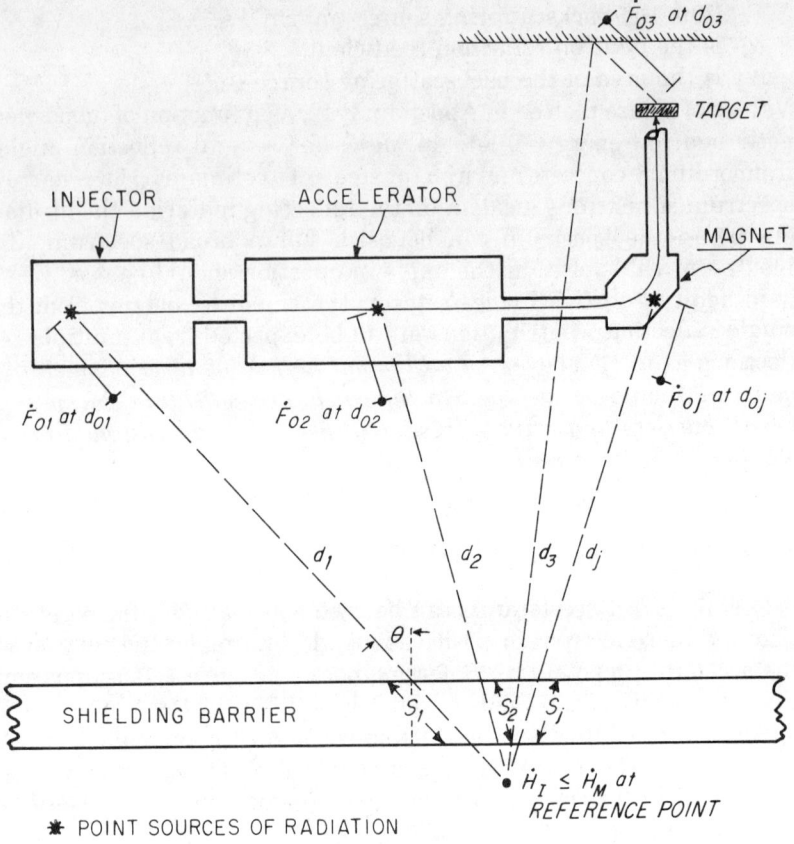

Fig. 5. Multiple sources of radiation. Diagram of a complex accelerator system illustrates the contribution to dose equivalent at a reference point outside the shielded area from j sources producing radiation simultaneously.

θ, where θ is the angle between the radiation direction and the normal to the wall face. For divergent beams, some judgment must be exercised in choosing the appropriate value of θ to derive the wall thickness.

4.4 Apertures

4.4.1 *General*

Shielding barriers are often interrupted by some kind of aperture, whether for personnel access, material flow, movement of apparatus, or for services to the accelerator facility such as ventilation, electric power and controls, water, and gas. Such apertures *shall* be designed so that the radiation protection of the interrupted shield *shall not* be impaired by insufficient radiation attenuation. Wherever practicable, apertures *should* be designed so that direct unattenuated radiations do not impinge on them.

Aperture designs are either in the form of mazes, in which advantage is taken of distance and the more readily attenuated reflected radiations, or in the form of heavy doors or plugs that are rolled or swung into place to seal the aperture. In selected cases, apertures can be surrounded by a fenced-off area beyond which the radiations are below H_M or dose-limit levels by virtue of the inverse-square relation. Occasionally, combinations of the above design concepts are utilized.

4.4.2 *Mazes and Ducts*

Depending on the nature of the radiation to be attenuated (and the ingenuity of the facility designer) mazes and large ducts can be designed to be both practical and safe. The major disadvantages of mazes for personnel access and movement of apparatus are that they are expensive, space consuming, and they often present problems in maneuvering apparatus or stretcher patients. Nevertheless, mazes are frequently to be preferred over radiation-protective doors, particularly in electron-accelerator or x-ray generator facilities where radiation energies are below 10 MeV (see (b) below).

a. *Mazes for electron accelerators.* The simplest method of shielding against exposure from electrons is to design a path length for the electrons that is greater than the electron range in air. This length can be in a straight line or the sum of the shortest mid-air distances in a maze. The length can be shortened by interposing a door at the maze entrance, with sufficient thickness to absorb the electrons, and with a seal, e.g. metallic weather-stripping, between door and aperture. This general design concept is useful for protection against scattered electrons that escape enclosed irradiation regions, or for low beam-power applications, e.g. electron therapy.

Since electrons can produce x rays wherever they impinge, protection *shall* be provided against the x rays in the maze design, in accordance with (b) below, for extended sources of x rays.

b. *Mazes for x-ray sources.* The design of a maze for x-ray attenuation (see Figure 6) can be calculated as a special case of x-ray reflection, described in Section 4.3.2. If at all possible, the inner aperture of an x-ray maze *should not* be exposed to the x rays that are emitted directly from the target, but rather *should* be exposed only to

62 / 4. RADIATION SHIELDING

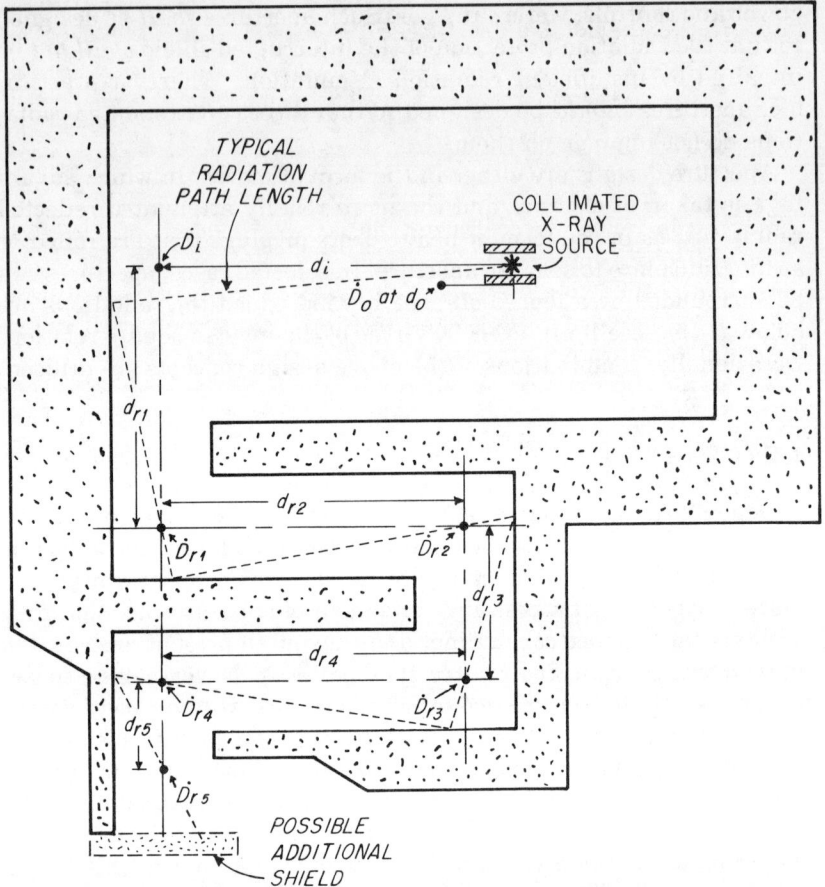

Fig. 6. Generalized maze design. This diagram illustrates successive reflections of x rays from a collimated source, through a maze. These path lengths can be approximated by a sequence of centerline distances, as shown in the diagram. In some instances, path lengths in very short legs can be omitted, e.g. d_{r3} or d_{r5}.

radiation reflected from shielding barriers or materials in the beam. Depending on the energy and emission rate of x rays incident on the first reflecting material, additional legs to the maze may be required to attenuate the radiation adequately. For economy in construction or space, it may be desirable to place an additional shielding baffle either before the inner aperture or outside the entrance into the maze, for additional attenuation and/or reflection of the radiation (see Figure 6).

For x rays below about 10 MeV in energy, the following calculation method provides a conservative estimate of the dose-equivalent index rate, $\dot{H}_{\mathrm{I,rj}}$, at the outside aperture of the maze:

$$\dot{H}_{\mathrm{I,rj}} \leq \dot{H}_{\mathrm{M}}$$
$$\dot{H}_{\mathrm{I,rj}} = \frac{\dot{D}_0\, \alpha_1\, A_1\, (\alpha_2\, A_2)^{j-1}}{(d_\mathrm{i}\cdot d_{\mathrm{r1}}\cdot d_{\mathrm{r2}}\cdots d_{\mathrm{rj}})^2} \quad (13)$$

where

- α_1 is the reflection coefficient for x rays incident on the first reflecting material;
- α_2 is the reflection coefficient for 0.5-MeV x rays reflected from subsequent materials (assumed to be the same for all subsequent reflection processes);
- A_1 is the area struck by x rays incident on the first reflecting material;
- A_2 is the cross-section of the maze (assumed to be approximately constant throughout the maze, with the height/width ratio between 1 and 2);
- $d_{\mathrm{r1}}, d_{\mathrm{r2}} \cdots d_{\mathrm{rj}}$ are the centerline distances along each maze length; the ratios $d_{\mathrm{rj}}/A_2^{1/2}$ should lie between 2 and 6;
- j refers to the jth reflection process.

Mazes for attenuating x rays above about 10 MeV in energy are more complex than described above, because of the necessity for thicker shielding barriers, uncertainties about reflection coefficients, the possibility of photoneutron production, and a significant component of annihilation radiation. Equation (13) is more conservative for incident x rays above 10 MeV. However, protective doors *should* be used at maze exits as an additional precaution, except for low-power radiotherapy installations.

In any maze design, the sum of the projected wall thicknesses in the maze, along a line from the x-ray source, *shall* be equivalent to the shielding-barrier thickness that otherwise would have been calculated if a maze had not been required on that wall.

c. *Mazes for neutron sources.* The matter of attenuating neutrons through mazes or ducts is a complex area of shielding technology. Because of this situation, shielded apertures in barriers for neutron-producing accelerators are often overdesigned. Occasionally, however, radiation surveying indicates a need for additional shielding. This uncertainty tends to lead facility designers toward the use of radiation-protective doors, rather than mazes.

Nonetheless, there is an empirical and conservative approach to the design of neutron mazes or large ducts. Reference is made to Maerker and Muckenthaler (1967).

The material of the shielding barriers is assumed to be (or have a reflection coefficient similar to) ordinary concrete. A large proportion of the neutrons within the accelerator room are those that have been reflected from all barriers surrounding the machine. These neutrons have a degraded spectrum with respect to the primary neutrons incident on the barriers. For accelerated-particle energies less than 100 MeV, the reflected-neutron spectrum increases markedly in fluence rate with decreasing energy. It is important to re-emphasize that the maximum intensity of the primary neutron beam *should not* impinge on the inner aperture of the maze or duct.

Fast neutrons tend to lose energy in scattering around passageway bends. The first leg of the maze, therefore, reduces primarily the fast-neutron fluence rate. In subsequent legs, thermal neutrons tend to establish dominance. For a multi-legged maze, it is fortuitously safe beyond the second leg to assume that *all* neutrons at the entrance have thermal energies. Gamma rays from the maze walls give about the same dose-equivalent index rate as the thermal neutrons. Thus, beyond the second leg, the total dose-equivalent index rate is about twice the thermal-neutron dose-equivalent index rate.

The behavior of the radiation can be expressed as a function of duct length, expressed in units of the square root of the duct cross section (i.e. $\sqrt{hw}$). The transmission factor for thermal neutrons through ducts is shown in Appendix F-11, for straight ducts as well as two- and three-legged ducts.

The neutron transmission ratio through mazes or ducts, B_{nm}, can be estimated as follows:

$$B_{nm} \leq \frac{270 \, \dot{H}_M}{K \, \phi_m} \qquad (14)$$

where
  ϕ_m is the neutron fluence rate incident on the inner aperture of the maze (cm^{-2} s^{-1});
  the value, 270, is the neutron fluence rate per unit dose-equivalent index rate for thermal neutrons (see Appendix B-3);
  K is a factor of conservatism, to include contributions to dose-equivalent index rate by gamma rays produced by neutron impingement on the walls of the maze, as well as to provide intentional overdesign of the maze (K = 8 for two-legged mazes; a value of 4 is adequate for three-legged mazes).

By means of the graph in Appendix F-11 the total maze length corresponding to the estimated B_{nm} can be obtained. This dimension is multiplied by $\sqrt{hw}$ to yield the maze length in feet.

4.4.3 *Shielded Doors*

There are many successful designs of shielded doors presently in use in accelerator facilities. Rather than to elaborate on the details of these designs, the purpose of this subsection is to outline the kinds of problems encountered in the design and use of shielded doors. Generally speaking, unless a maze entrance way is used, a door into a facility must close the aperture with a structure that is equivalent in shielding thickness to the wall surrounding the aperture. This door must be moved by some reliable method away from the aperture to permit unobstructed entry into the facility. The closing of the door must be accomplished without hazard to personnel, but with certainty that the aperture is completely blocked with adequate shielding material. A shielded door is necessarily heavy and bulky; the translation of this mass is mainly a mechanical engineering problem.

Materials most commonly used in door construction include ordinary concrete, steel, or lead. The latter two materials are used mainly to conserve space. Because lead tends to flow under its own weight, it is usually mounted on a steel structure or is sandwiched between steel plates. Concrete can be cast into a supporting structure, but more often it is stacked and mortared in the form of blocks or bricks. In either case, it is important to eliminate voids in the shield.

Steel and lead doors can be mounted on hinges to be swung open, or they can be suspended from a roller track to be moved laterally. Concrete doors are usually mounted on wheels or rollers, guided by a track system. To block openings in very thick shielding walls, concrete plugs are occasionally used. These are rolled away from the wall into the area outside the radiation room. Another type of shielded door that can be incorporated into a thick concrete barrier is in the form of a vertical cylinder with an aperture that can be rotated for access into the radiation area. When the door is properly closed, the shielding on either side of the aperture provides protection equivalent to the adjacent barrier.

The major shielding problem to be solved in a door design is that of providing sufficient overlap of the door over the aperture, either to minimize radiation reflection through the gap between door and wall, or to avoid penetration of direct radiation through insufficient material, i.e. by undercutting. This problem can be readily solved at the top and sides of the door by providing an overlap at least 10 times the gap between door and wall (see Figure 7). The door may be mounted inside the radiation room to minimize edge-penetration problems. A

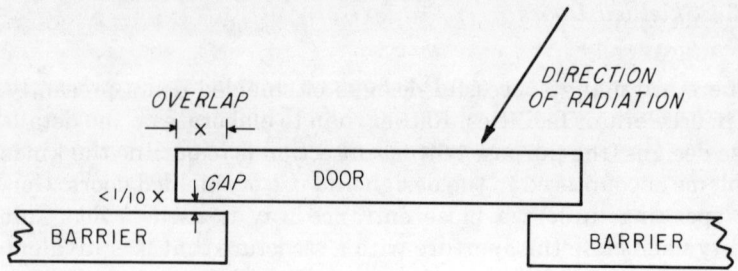

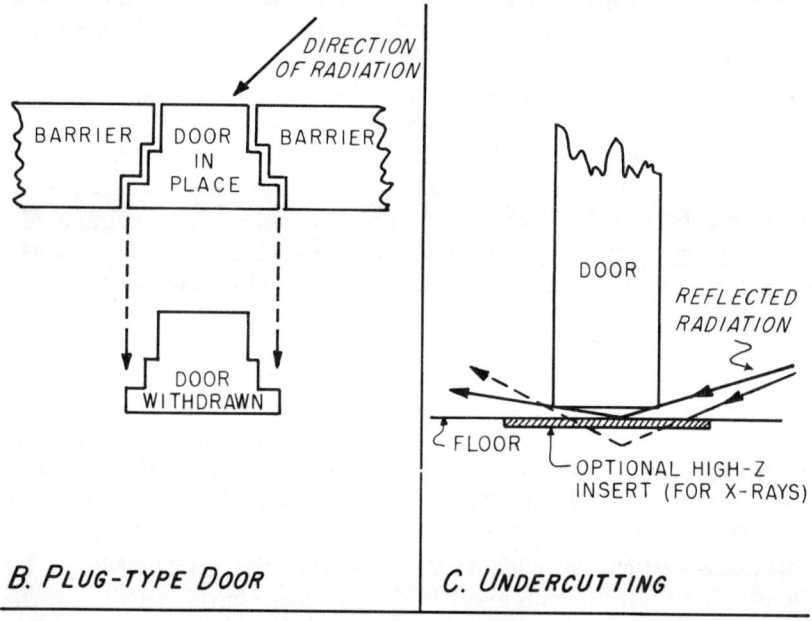

Fig. 7. Details of typical shielded doors.

concrete plug may be stepped or tapered to offset the gap with respect to direct radiation.

The bottom of the door presents a special problem in radiation undercutting or reflection, because of the necessity of providing a smooth floor when the door is open, for moving equipment or patients and for personnel safety. If the door is recessed into the floor, a flush threshold must be provided to fill or cover the slot in the floor. If the closed door is mounted, e.g. on wheels or rollers, so that a gap exists at floor level, there may be a problem with radiation penetrating and reflecting through the material in the floor, as well as the radiation reflecting through the gap itself (see Figure 7C). Some additional means of closing the gap may be required for the latter problem, such as settling the door into place or raising the threshold to meet the door. The floor material under the door should be selected to attenuate adequately the radiation leaking through the shortest path under the door.

Heavy shielded doors are driven by electrical, hydraulic, or pneumatic means. In any case, they constitute a hazard for personnel, particularly in the closing mode of operation, and they should be interlocked with an elevator-type door safety edge, or a "dead-man" switch, or some other means by which the motion of the door can be stopped in an emergency. In addition, a mechanical system *should* be provided to operate the door manually in the event of power failure.

4.4.4 *Openings for Material Flow*

The following kinds of openings in shielding barriers are considered in this subsection: pipes and ducts for continuous flow of homogeneous material; tunnels and ducts for intermittent flow of heterogeneous material; pipes for conveying samples intermittently in and out of the radiation area. In all of these cases, the design of the opening *shall* provide adequate shielding against the radiations produced inside the facility, at least to an extent equivalent to the adjacent shielding barrier.

a. *Pipes and ducts for continuous flow of homogeneous material* include water and gas pipes for servicing the accelerator as well as pipes and ducts for conveying fluidized material into and out of the facility, e.g. for electron-beam processing. Such pipes have cross-sections that are small in comparison with the dimensions of the shielding barriers. The orientation of the pipes through the shielding barrier *should* be designed to avoid alignment of the pipe centerline with the direct radiation. The total shielding thickness *should* be calculated with consideration of the small void in the shielding made by the projected cross-section of the pipe and the material continuously flowing through it. The bends in the pipe *should not* be abrupt, so that hazards of clogging within the shielding barrier are minimized.

b. *Tunnels and ducts for intermittent flow of heterogeneous material* can be considered as special cases of mazes and ducts considered in Section 4.4.2. A tunnel can be designed in the form of a small maze, for conveying packaged goods or materials in and out of the radiation area. The cross-section of the tunnel *should* be no greater than required for trouble-free conveyance of the material. For electron-processing purposes, the total path length through the tunnel to the electron beam *should* be equivalent to, or greater than, the maximum range of the electrons through air. If this is not practical, e.g. at electron energies above a few MeV, some additional baffling *should* be designed into the tunnel or its conveyor, to absorb the electrons. In some cases, the material itself can serve as such a baffle. Interstitial baffles between packages can be designed to provide more complete absorption. In any case, the maze *shall* be designed to attenuate the x-radiation produced by electron impingement on the conveyed material. Since most irradiated materials are low-Z in nature, x-ray production would be minimized. It is important, however, to line the conveyor system with low-Z material in the immediate vicinity of the electron beam, to minimize x-ray production when material is not being conveyed.

Flexible materials such as plastic films, textiles, and paper goods can be fed into the radiation area of an electron-processing accelerator through slots that are angled sharply with respect to the electron-beam direction. If necessary, these materials can be flexed in direction prior to entry into the shielding barrier, behind a supplementary shield between the entrance to the slot and personnel in the area.

c. *Pipes for conveying samples,* e.g. pneumatic "rabbit" systems for isotope production and activation analysis, *should* be designed into shielding barriers as described in (a) above. Because of the likelihood that radioactive material is to be transferred through such a system, the pipe *should* be designed to avoid jamming of the rabbit. The bends in the pipe or conduit *should* be gradual to facilitate clearing the line.

4.5 Special Shielding Problems

4.5.1 *Skyshine*

In the design of accelerator facilities that are enclosed in a separate building located at some distance from other buildings, the question often arises concerning the magnitude of shielding that is required for the roof over the radiation area. Assuming that an ordinary roof, i.e. one designed only for protection against weather, provides little if any attenuation for upward-directed radiation from the accelerator, there is a significant probability that radiation reflected back from the atmosphere will exceed H_M or dose-limit levels in the immediate area of the facility. This reflected radiation is termed "skyshine." Reference is made to Figure 8.

An estimate of the absorbed-dose index rate from *x-ray* skyshine can be obtained from the following equation (adapted from Clarke, 1968):

$$\dot{D}_{Is,d_s} = [(2.5 \times 10^{-2})\, \dot{D}_{I0}\, \Omega^{1.3}]/d_s^2 \qquad (15)$$

where

$\dot{D}_{Is,d_s}$ is the absorbed-dose index rate from x-ray skyshine at distance d_s (rads m² min⁻¹);

Ω is the solid angle (measured in steradians) subtended by the x-ray source and the shielding walls (see Figure 8).

This relationship provides a conservative estimate for distances between 20 and 250 meters and for relatively low photon energies. It is increasingly on the safe side for higher energy photons and for shorter distances.

The shield thickness for the roof can be calculated by the method described in Section 4.3.2. When skyshine dominates the shielding situation, the roof shielding transmission ratio, B_{xs}, is obtainable through a modification of Equation (3), replacing $\dot{D}_{I0}$ of that equation by $\dot{D}_{Is,d_s}$ of Equation (15), as follows:

$$B_{xs} \leq (0.67 \times 10^{-3}) \left[\frac{\dot{H}_M\, d_i^2\, d_s^2}{\dot{D}_{I0}\, \Omega^{1.3}} \right] \qquad (16)$$

where

d_i is the distance between x-ray source and about 2 meters above the roof;

$\dot{D}_{I0}$ is the absorbed-dose index rate (rads m² min⁻¹), measured in the upward direction from the x-ray source, at a standard reference distance of 1 meter.

The resulting shield thickness may alternatively be designed into the roof structure or mounted directly over the x-ray source with a lateral area sufficient to cover the solid angle, Ω. The latter design requires less shielding material and structural support for the same solid-angle coverage; however, Section 4.5.2 below *should* be reviewed with regard for precautions to be taken with such a design.

Skyshine from *neutrons* can be estimated in a manner analogous to the x-ray case. The following equations were derived by Chilton (1974) from the work of Ladu *et al.* (1968):

$$\phi_{s,d_s} \simeq (6.5 \times 10^{-2})\, (\phi_0\, \Omega)/(2\pi\, d_s^{1.6}) \quad \text{(for } d_s > 20 \text{ meters)}; \qquad (17)$$

$$\simeq (5.4 \times 10^{-4})\, (\phi_0\, \Omega)/2\pi \quad \text{(for } d_s \leq 20 \text{ meters)}. \qquad (18)$$

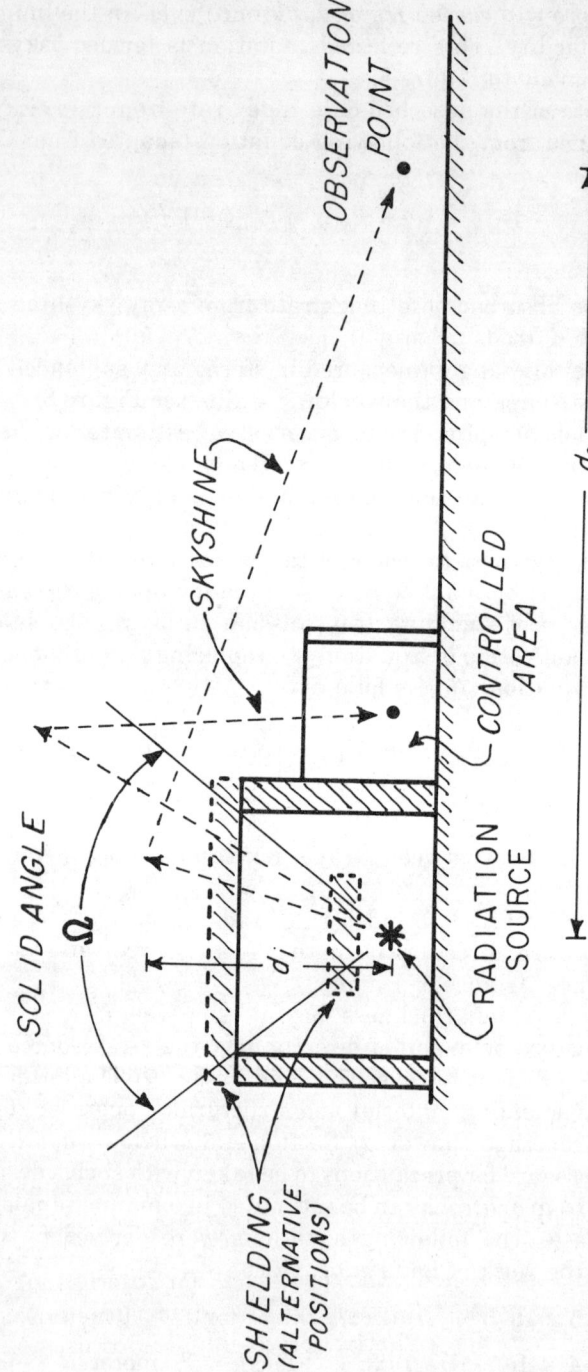

Fig. 8. Geometrical considerations for skyshine.

where ϕ_{s,d_s} is the skyshine-neutron fluence rate (m² cm⁻² s⁻¹) at distance d_s.

These equations relate to a point isotropic source of 5-MeV neutrons, but they provide a reasonable estimate of skyshine from neutrons of other energies. For example, Lindenbaum (1957) estimated that the skyshine-neutron fluence rate is about 10 percent of the unshielded rate at 12 meters from the neutron source, for neutrons in the 1- to 10-MeV range, assuming $\Omega/2 \sim 1$ (i.e. the open upper half plane), and that the barrier to the sideward-directed radiation is "infinitely" thick. This estimate is about 30 percent higher than that obtained from Equation (18). For purposes of calculating shielding, the values obtained from Equations (17) and (18) *should* be increased by a factor of two, to be conservative.

The shield thickness for the roof can be calculated by the method described in Section 4.3.3. When neutron skyshine dominates the shielding situation, the roof shielding transmission ratio, B_{ns}, is obtainable through a modification of Equation (11), replacing ϕ_0 of that equation by ϕ_{s,d_s} of Equation (17) or (18), as follows:

$$B_{ns} \leq (2.7 \times 10^{-5}) \left[\frac{\dot{H}_M d_i^2 d_s^{1.6}}{\phi_0 \Omega} \right] \quad \text{(for } d_s > 20 \text{ meters)}; \qquad (19)$$

$$\leq (3.3 \times 10^{-3}) \left[\frac{\dot{H}_M d_i^2}{\phi_0 \Omega} \right] \quad \text{(for } d_s \leq 20 \text{ meters)}. \qquad (20)$$

This calculation method is expected to be increasingly safe for neutrons of energy greater than 10 MeV.

4.5.2 Detachable and Temporary Shielding

There are many reasons for utilizing shielding barriers that can be removed or that are installed for temporary purposes. Unless the facility has been designed for safe operation of the accelerator *without* such shields, however, there are hazards attending their use. A few examples follow:

Certain low-energy ion accelerators require thin lead barriers to shield against the x rays produced by back-streaming electrons. These barriers are occasionally designed in the form of movable structures which are placed close to the x-ray source except during servicing. The proper positions and configurations of these forms *shall* be firmly established, with provisions for interlocking them with the accelerator control system. In the same sense, low-energy electron accelerators for radiation processing are occasionally installed with movable shielding forms that directly surround the machine and its irradiation volume. These forms *shall* be designed and interlocked so as to require that they be placed into their safe configuration.

In some therapy x-ray generator designs of the isocentric type the x-ray beam stopper that is placed directly opposite the x-ray head can be moved out of the way for special treatment purposes. The shielding barriers that can be exposed to the unattenuated useful beam *shall* be calculated on the basis that the beam stopper is removed.

Direct accelerators that are used in research can be used alternatively for the acceleration of ions or electrons, as described in Section 4.3.4. The shielding requirements differ for these two modes of operation, and the local-shielding needs may also differ. The facility *should* be designed for the worst radiation case, assuming no local shielding unless such shielding can be unambiguously mounted and interlocked into position.

Barrier chains and stanchions are frequently used to restrict personnel from a high radiation area. They *shall* be interlocked and their proper position *shall* be clearly designated.

4.5.3 *Computer Programs for Calculating Shielding Thickness*

There are a number of computer codes that are available for more exact methods of calculating shielding thickness. These are available from many different sources, including Code Centers in various National Laboratories of the Energy Research and Development Administration. A particularly appropriate source would be in the Radiation Shielding Information Center, at the Oak Ridge National Laboratory, Oak Ridge, Tennessee. Information on codes is also available in the literature. The selection and use of codes *should*, however, be made only with the guidance of a shielding specialist.

APPENDIX A

Definitions

A.1 Definition of Terms

The following definitions are given for purposes of clarifying the contents of this report. In some instances, they may differ somewhat from common usage. Many of the quantities and units defined below have been the subject of extensive analysis by the International Commission on Radiation Units and Measurements (ICRU). More precise definitions of these quantities and units may be found in ICRU Report 19 (ICRU, 1971a) and ICRU Report 19S (ICRU, 1973).

absorbed dose (D): The quotient of $d\bar{E}$ by dm, where $d\bar{E}$ is the mean energy imparted by ionizing radiation to the matter in a volume element and dm is the mass of the matter in that volume element. The special unit of absorbed dose is the *rad*.

absorbed dose index (D_1): The maximum absorbed dose within a 30-cm diameter sphere centered at the point of interest and consisting of material equivalent to soft tissue with a density of 1 g cm^{-3}.

absorption: A phenomenon in which some or all of the energy in a beam of radiation is transferred to the matter which it traverses.

accelerator (see also electron-beam generator, neutron generator, x-ray generator): A device for imparting kinetic energy to charged particles. For this report, the energy is in general greater than 0.1 MeV.

activation: The process of inducing radioactivity by irradiation.

activity: The number of spontaneous nuclear transformations which occur in a quantity of a radioactive nuclide per unit time. The special unit of activity is the *curie*.

albedo (see also reflection coefficient): The ratio of the absorbed dose index or particle fluence reflected out of a non-source medium to the absorbed dose index or particle fluence into it.

anisotropic (see also isotropic): Not isotropic; having different properties in different directions.

annihilation radiation: The electromagnetic radiation emitted as a result of the combination and disappearance of an electron and a positron. Two gamma rays of 0.511 MeV energy each are emitted in most cases.

area (see controlled, exclusion, noncontrolled, occupiable, radiation, radioactivity areas).

area occupancy factor (T): A factor (≤ 1) by which the absorbed dose index rate or neutron fluence rate, used for shielding calculations, should be multiplied to correct for the degree of occupancy of the area in question while the radiation source is "on" and the barrier protecting the point of interest is being irradiated.

attenuation: The reduction of dose-equivalent index rate upon passage of radiation through matter. This report is concerned with broad-beam attenuation, i.e. that occurring when the radiation field is wide at the barrier.

attenuation length: The penetration distance in which a radiation beam is attenuated by a factor of e.

backreflection (scattering): The reflection of radiation in a direction generally greater than 90° to that of the incident radiation.

beam (also charged-particle beam, radiation beam): A flow of electromagnetic or particulate radiation that is either (a) collimated and generally unidirectional, or (b) divergent from a small source but restricted to a small solid angle.

bremsstrahlung (see also x ray): The electromagnetic radiation associated with the acceleration or deceleration of charged particles, particularly in the vicinity of the Coulomb fields of nuclei.

broad-beam conditions: Conditions of a radiation-shielding situation in which the beam impinging on the shield surface includes scattered radiation and is laterally extensive.

buildup (of radiation in shield): That part of the total value of a specified radiation quantity at any point due to radiation that has undergone interactions in the shield or that results from such interactions.

capture gamma ray: A photon emitted as an immediate result of the neutron-capture process.

characteristic x ray: X rays which are characteristic of the element in which they are produced. Their emission results from the rearrangement of electrons in the shells of excited atoms.

charged particle: An atomic or subatomic quantity of matter (e.g. electron, proton, alpha particle, ionized atom) possessing or lacking a net electrical charge of one or more elementary units of charge.

collimate: To reduce the cross-sectional area of a divergent beam of photons or particles.

collimator: Any arrangement of slits or apertures which limits a stream of particles or photons to a beam in which all particles or photons move in the same (or nearly the same) direction.

Compton scattering: The elastic scattering of a photon by an essentially free electron.

controlled area: A defined area in which the exposure of persons to radiation or to radioactive material is under the supervision of a radiation-protection (-safety) supervisor. (This definition implies that a controlled area is one that requires control of access, occupancy, and working conditions for radiation protection purposes.)

curie (Ci): The special unit of activity. One curie is exactly 3.7×10^{10} nuclear transformations per second (s^{-1}). Also, the quantity of any radioactive material having an activity of one curie.

dark current: A current, usually of electrons, that may flow through an acceleration tube or waveguide from sources other than the cathode of the accelerator. This is an abnormal phenomenon, often associated with poor vacuum conditions or contaminated surfaces in the acceleration region.

dead-man switch: A switch so constructed that it remains activated only by continuous pressure on the switch.

diaphragm: A device with a central aperture so designed as to restrict the beam to an appropriate area at the point of interest.

direct radiation (see also primary radiation): Radiation emitted rectilinearly from the target or source.

directly ionizing radiation: Charged particles (electrons, protons, alpha particles, etc.) having sufficient kinetic energy to produce ionization by collision.

dose: A colloquial term. For precision, it should be appropriately qualified (see absorbed dose, absorbed dose index, dose equivalent, dose-equivalent index).

dose equivalent (H): The product of absorbed dose, D, quality factor, Q, and N, at the point of interest in tissue. N is the product of any modifying factors other than Q. For the purposes of this report, N = 1. The special unit of dose equivalent is the *rem*. With D expressed in *rads*, H is in *rems*. The dose equivalent is not defined for values larger than the order of the maximum permissible dose equivalent, H_M. There is no such restriction on dose-equivalent rate, $\dot{H}$.

dose-equivalent index (H_I): The maximum dose equivalent within a 30-cm diameter sphere centered at the point of interest and consisting of material equivalent to soft tissue with a density of 1 g cm^{-3}. In general, the maximum values of the quality factor, Q, and the absorbed dose, D, occur at different locations in the sphere. However, $Q_{max}D_{max}$ is a conservative estimate for H_I.

dose limit: For radiation protection purposes, the maximum dose equivalent that the general public shall be allowed to receive in a stated period of time (see Appendix B; also maximum permissible dose equivalent).

drift-tube (type of accelerator): A series of hollow cylindrical conductors in a linear particle accelerator, with potential differences applied between successive tubes. The charged particles travel through the field-free drift tubes with uniform speed and are accelerated only at the gaps between successive tubes.

duty cycle: The fraction of the operation cycle of an accelerator during which radiation is produced; the product of the pulse duration and the pulse-repetition frequency.

efficiency (of x-ray production): The fraction of electron power incident on a target that is converted to x-ray power.

electron-beam generator: A type of electron accelerator in which the electron beam is brought out into the atmosphere for irradiation purposes.

endoergic; endothermal: Characterized by the absorption of energy or heat. Endoergic reactions absorb energy as they progress. Endothermal reactions absorb heat as they progress.

energy, low or high: In this report, the kinetic energy of particles or photons. Low energy is considered as less than 10 MeV, and high energy is considered as greater than 10 MeV.

equilibrium tenth-value layer (or half-value layer): The thickness of a specific material that attenuates a specified radiation by a factor of ten (or two), under broad-beam conditions, in that penetration region where the radiation directional and spectral distributions are practically independent of thickness, so that a single value of the tenth-value layer (or half-value layer) is valid.

exoergic; exothermal: Characterized by the production of energy or heat.

exclusion area: An area defined by the radiation protection officer to be forbidden to all personnel during operation of the accelerator.

exposure (X): A measure of x- or gamma-ray radiation based upon the ionization produced in air by x rays or gamma rays. The special unit of exposure is the roentgen (R).

extended source (of radiation): A source of particles or photons which cannot be considered a point source, e.g. whose linear dimensions are greater than 1/10 the distance between source and observation point.

face, entrance (of shielding barrier): The surface of the shielding barrier on which radiation is incident.

face, exit: The surface of the shielding barrier from which attenuated radiation leaves the shielding material.

fail-safe (relating to circuit or system): Having the property that any failure causes a sequence of actions which *always* results in a safe situation.

fast neutrons: Neutrons of energies above a few hundred keV.

fluence, particle (Φ): The quotient of dN by da, where dN is the number of particles which enter a sphere of cross-sectional area da. Φ is usually expressed in cm^{-2}.

fluence rate, particle; flux density (ϕ): The quotient of $d\Phi$ by dt, where $d\Phi$ is the increment of particle fluence in the time interval dt. ϕ is usually expressed in cm^{-2} s^{-1}.

gamma ray. Electromagnetic radiation emitted in the process of nuclear transition or radioactive decay.

general public: In the context of this report, the general mass of the populace not regarded as radiation workers.

half-life, radioactive: The time for the activity of any particular radioactive nuclide to be reduced to one-half its initial value.

half-value layer (or thickness): The thickness of a specified substance which, when introduced into the path of a given beam of radiation, reduces the absorbed dose index or dose-equivalent index by one-half. The magnitude of the half-value layer may be different for absorbed dose index and dose-equivalent index.

high radiation area: Any area, accessible to personnel, in which there exists radiation at such levels that a major portion of the body could receive in any one hour a dose equivalent, H, in excess of 100 mrem.

indirectly ionizing radiation: Uncharged particles (neutrons, photons, etc.) which can liberate directly ionizing particles or can cause nuclear transformations.

interference (in shielding barrier): Discontinuity or void in a shielding

APPENDIX A / 77

barrier, e.g. aperture, piping, ductwork, maze, which tends to reduce the effective thickness of the barrier.

interlock: Device which automatically shuts off or reduces the radiation emission rate from an accelerator to acceptable levels, e.g., by the opening of a door into a radiation area. In certain applications, an interlock can be used to prevent entry into a radiation area.

intermediate neutrons; slow neutrons: Neutrons with energies of about 1 eV to a few hundred keV.

inverse square relation: That rule which states that the intensity of radiation from a point source decreases as $1/d^2$ from the source in a non-absorbing medium, where d is the distance from the source.

ionizing radiation: Any radiation consisting of directly or indirectly ionizing particles or photons or a mixture of both.

irradiation: Exposure to ionizing radiation.

isotropic (see also anisotropic): A condition in which properties are the same in whatever direction they are measured. With reference to radiation emission, this term indicates equal emission in all directions from a point source or each differential-size element of any extended source.

kerma: The sum of the initial kinetic energies of all the charged particles liberated by indirectly ionizing particles per unit mass of specified material. The special unit of kerma is the rad.

leakage radiation: All radiation, except the useful beam, coming from within the accelerator components, e.g., that radiation that is attenuated by a collimator, diaphragm, or source shielding.

linear energy transfer (LET): The average energy lost by a directly ionizing particle per unit distance of its travel in a medium.

maximum permissible dose equivalent (H_M): For radiation protection purposes, H_M refers to the maximum dose equivalent that radiation workers *shall* be allowed to receive in a stated period of time (see Appendix B; also dose limit).

mean free path (for a given type of interaction, e.g., scattering or absorption): The average distance that particles of a specified type travel before a specified type of interaction takes place in a given medium. If the term *mean free path* is used without specifying the interaction, the term means the average distance a particle will travel before having an interaction of any sort.

monitor, radiation: A radiation-measuring assembly provided with devices intended to draw attention to an event or situation which might result in overexposure to the radiation. The assembly may include indicating and/or recording instruments.

monoenergetic: Possessing a single energy. This term is sometimes used to characterize a radiation field in which the particles or photons have various energies within a narrowly limited band.

narrow-beam conditions: Conditions of a radiation-shielding situation in which the beam impinging on the shield surface excludes scattered radiation and is laterally restricted.

neutron capture: A process in which a neutron becomes part of the nucleus with which it collides without release of another heavy particle.

neutron generator: A type of accelerator in which the ion beam or the x-ray beam is used mainly for the production of neutrons; e.g. low-voltage deuteron accelerators.

noncontrolled area: Any space not meeting the definition of controlled area.

nonoccupational exposure: Radiation exposure received by an individual that is not expected as part of his normal occupation.

nuclear reaction: An interaction between a photon, particle or nucleus and a target nucleus, leading to the emission of one or more particles and/or photons.

nuclide: A species of atom having specified numbers of neutrons and protons in its nucleus.

occupational exposure: Exposure of an individual to ionizing radiation in a controlled area or in the course of employment in which the individual's normal duties or authorized activities necessarily involve the likelihood of exposure to ionizing radiation.

occupiable area: Any room or other space, indoors or outdoors, that is likely to be occupied during an irradiation, by any person, either regularly or periodically during the course of his work, habitation, or recreation.

ordinary concrete: A Portland-cement concrete whose constituents are those usually utilized in construction. Thus, ordinary concrete excludes those mixtures where special materials (iron, etc.) have been added to enhance the radiation-shielding properties. For example, the term excludes "heavy concrete."

pair production: The simultaneous production of an electron and a positron by an interaction of a photon or a fast charged particle with the electronic field of a nucleus or other particle.

path length: Total length of the path of a particle moving through a medium, measured along the actual path, whether or not rectilinear.

photoneutron: A neutron released from a nucleus by a photonuclear reaction.

photodisintegration: Any transformation of a nucleus induced by photons.

photonuclear: Pertaining to the interaction of a photon with a nucleus.

photofission: Fission of a nucleus induced by a photon.

photoelectric effect: The interaction of a photon with an atom, resulting in the absorption of the incident photon and the release of a bound electron from that atom with energy equal to the photon energy less the electron binding energy.

photon: An energy quantum of electromagnetic radiation. In this report, an x-ray or gamma-ray photon.

point source (of radiation): Any radiation source measured from a distance that is much greater than the linear size of the source. In this report, a source whose linear dimensions are less than 10 percent of the measurement distance may be considered a point source for shielding purposes.

polyenergetic: Possessing many different energies, either in a continuous spectrum or in groups of energy bands.

primary radiation: The radiation of earliest origin in the context of the situation considered. For example, electron radiation impinging on a shielding wall may be considered as "primary," from the standpoint of shielding against electrons. The x rays produced from this impingement may be considered as "primary" from the standpoint of shielding against x rays.

qualified expert on radiation shielding: A person having the knowledge and training to undertake the analysis and design of a radiation-shielding system.

qualified health physicist: A person professionally engaged in radiation protection, who is certified by the American Board of Health Physics or who has equivalent competence.

qualified radiological physicist: A person professionally engaged in the physics applied to radiology, who is certified in this field by the American Board of Radiology or who has equivalent competence.

quality factor (Q): A factor which is used in radiation protection to weight the absorbed dose with regard to its presumed biological effectiveness insofar as it depends on the LET of the charged particles. The quality factor is a function of the LET of the charged particles that deliver the absorbed dose.

rabbit: A small sample container propelled pneumatically through a tube leading from the laboratory to a location in a nuclear reactor or near an accelerator where irradiation can take place, and designed to provide short transit times back to the laboratory.

rad: The special unit of absorbed dose. 1 rad equals 10^{-2} joules kg^{-1}, or 100 ergs g^{-1} (plural: rads).

radiation: Propagation of energy through space. In the context of this report, electromagnetic radiation (x rays or gamma rays), or corpuscular radiation (electrons, protons, atomic ions, neutrons, heavy particles), capable of producing ionization.

radiation area: Any area, accessible to personnel, in which there may exist radiation at such levels that a major portion of the body could receive in any one hour a dose equivalent, H, in excess of 2.5 mrem, or in any 5 consecutive days a dose equivalent in excess of 100 mrem.

radiation field map: A drawing of a region traversed by radiation of any kind, showing contours of iso-absorbed dose index rate, isodose-equivalent rate, or iso-fluence rate; sometimes referred to as "radiation profile."

radiation length: The mean path length required to reduce the energy of a relativistic charged particle by a factor of e.

radiation level: The dose-equivalent rate of the radiation field at the point in question.

radiation protection (safety) officer: The person directly responsible for radiation protection.

radiation (protection) survey: An evaluation of the radiation safety in and around an installation.

radiation worker: One who works with or around radiation, or who, in the course of completing a task, may be exposed to radiation (e.g. an x-ray technician).

radioactive contamination: Radioactive substance dispersed in materials or places where it is undesirable.

radioactivity, induced: Radioactivity in nuclides produced by nuclear reactions.

radioactivity area: Any area in which radioactive materials are present.

range, ion or electron: The distance into a material that a charged particle penetrates before it ceases to ionize. "Projected or extrapolated range" is the distance into a material that is determined by extrapolating the straight-line portion of the curve of absorbed dose vs. distance, to zero dose. "Maximum range" is the distance into a material that is determined by the interception of the curve of absorbed dose vs. distance with the radiation background; used particularly to determine the maximum range of electrons.

redundancy (in interlock systems): Repetition; a situation in which two or more systems are designed to perform the same or approximately the same function, thus providing a safety factor in the instance of the failure of one of the systems.

reflected (scattered) radiation: Radiation that, during passage through matter, has been deviated in direction. It may have been modified also by a change, usually a decrease, in energy.

reflection coefficient: The ratio of the absorbed dose index rate (dose-equivalent index rate) of reflected radiation at 1 meter from the scatterer divided by the product of the absorbed dose index rate (dose-equivalent index rate) incident on the scatterer and the area of the scatterer that is irradiated.

relative biological effectiveness (RBE): Biological potency of one radiation as compared with another, in terms of the inverse ratio of the respective absorbed doses that produce the same biological effect. The use of this term is to be restricted to radiobiology, and it should be distinguished from the quality factor, Q, which is employed in radiation protection.

rem: The special unit of dose equivalent.

roentgen (R): The special unit of exposure equal to 2.58×10^{-4} coulomb per kilogram (plural: roentgens).

secondary electrons: Electrons ejected from an atom, molecule, or surface as a result of impingement by a charged particle or a photon.

secondary radiation: Particles or photons produced by the interaction with matter of a type of radiation regarded as "primary."

self-shielding: In accelerator practice, characteristic of a radiation-source design in which sufficient shielding material is incorporated adjacent to the source to reduce external dose rates below H_M levels or dose limits.

shall: Indicative of a recommendation that is necessary to meet the currently accepted standards of radiation protection.

shielding, local: Shielding material installed adjacent to, or close by, a radiation source, e.g. diaphragm and collimator around x-ray-producing target.

shielding transmission ratio (for x rays or neutrons): The ratio of the detector response at a location behind a shield on which radiation is incident to the detector response at the same location without the presence of the shield; a measure of the effectiveness of the shield.

should: Indicative of a recommendation that is to be applied when practicable.

skyshine: Radiation reflected back to earth by the atmosphere above a radiation-producing facility.

slowing down (of neutrons): Decrease in energy of a neutron, usually due to repetitive collisions of the neutrons with the matter through which they traverse.

special unit: A unit reserved for specified quantities only. Other units may also be employed for the same quantity. Thus, the rad is the special unit of absorbed dose (or kerma) which may also be expressed in ergs g^{-1}, $J\ kg^{-1}$, etc.

stopping power (of electrons or ions): A measure of the average energy loss of a charged particle passing through a material. Linear stopping power is specified as energy loss per unit distance traveled. Mass stopping power is specified as energy lost per unit distance traveled, divided by the density of the material.

straggling (of electrons or ions): The random variation or fluctuation of a property associated with charged particles in passing through matter. It is applied especially to range or penetration distance.

target: Any object or surface of an object bombarded by particles or photons.

tenth-value layer (or thickness): The thickness of a specified substance which, when introduced into the path of a given beam of radiation, reduces the absorbed dose index or dose-equivalent index to one-tenth. The magnitude of the tenth-value layer may be different for absorbed dose index and dose equivalent index.

thermal neutrons: Neutrons in thermal equilibrium with their surroundings. In this report, all neutrons with energies of less than about 1 eV are termed "thermal."

thick shield: Shield which effects a substantial reduction of a radiation field as a result of its presence. For the purposes of this report, the term usually implies a shielding transmission ratio of 10^{-6} or less.

thick target: Target whose dimension in the direction of incident particulate radiation is equal to or greater than the range of the incident particles.

threshold, nuclear-reaction: The minimum particle or photon energy required to initiate a specific endothermal (endoergic) nuclear reaction.

threshold, radiation-effect (or radiation-damage): The minimum absorbed dose (or dose equivalent) of radiation that will produce a specified effect or a specified type of damage to the irradiated material.

transmission ratio (see shielding transmission ratio).

undercutting: Penetration of radiation through cracks of shielding barriers or through thin sections of such barriers (e.g. edges of structures), resulting in a greater dose-equivalent rate than that resulting from passage of radiation through the bulk of the shielding barrier.

useful beam (or radiation): That part of the radiation from a target which emerges from the source and its housing through an aperture, diaphragm, or collimator.

week, calendar: 7 consecutive days.

week, work: Any combination of time intervals adding up to 40 hours within 7 consecutive days.

workload (W): The degree of use of an x-ray or gamma-ray source. For therapy x-ray machines, W is usually expressed in terms of mA min week^{-1} (below 4 MeV) and rem m^2 week^{-1} (above 4 MeV).

x ray: Electromagnetic radiation of energy usually above 1 keV. X rays are produced by impingement of charged particles (usually electrons) on materials (see bremsstrahlung and characteristic x rays).

x-ray generator: A type of electron accelerator in which the electron beam is used mainly for the production of x rays.

x-ray converter: Material in which electron energy is converted to x-ray energy, e.g. a thick target of high-Z material. In this report, the term is usually applied to a target in which electron power is converted with a high degree of efficiency into x-ray power.

yield (Y): In the context of this report, the total radiation emitted per unit time from an accelerator target as measured over the entire solid angle of 4π, divided by the beam current incident on the target.

Z; low-Z, high-Z: The symbol for the atomic number of a nucleus, i.e. the number of protons contained in the nucleus. Low-Z describes nuclei with $Z \leq 26$. High-Z describes nuclei with $Z > 26$. Very high-Z describes nuclei with $Z > 73$.

A.2 Definitions of Symbols

Symbol	Definition
A	area
A	ampere
B	shielding transmission factor
C	coulomb
Ci	curie
cm	centimeter
$D; \dot{D}$	absorbed dose; absorbed dose rate
$D_I; \dot{D}_I$	absorbed dose index; absorbed dose index rate
d	distance
d	deuteron
E	energy of particulate or electromagnetic radiation
e	elementary electronic charge
$F; \dot{F}$	radiation emission; radiation emission rate from a source (D_I, $\dot{D}_I$ for x rays; Φ, ϕ for neutrons)
f	pump exhaust rate
G	molecules formed per 100 eV energy absorbed
G	giga- (10^9)
g	gram
$H; \dot{H}$	dose equivalent; dose-equivalent rate
$H_I; \dot{H}_I$	dose-equivalent index; dose-equivalent index rate
$H_M; \dot{H}_M$	maximum permissible dose equivalent; maximum permissible dose-equivalent rate; (with additional subscript, H_{Mt} refers to the unit used in radiotherapy operations)
h	aperture height
h	hour
I	current of charged particles, expressed in amperes or submultiples thereof
i	as a subscript, incident
J	joule
j	last (jth) parameter in series
k	kilo- (10^3)
LET	linear energy transfer
M	mega- (10^6)
m	milli- (10^{-3})
m	meter
N	age in years
N	modifying factor for dose equivalent
n	nano- (10^{-9})
n	neutron
o	as a subscript, standard reference distance
p	pico- (10^{-12})
p	proton
$Q; \bar{Q}$	quality factor; mean quality factor
R	roentgen

Symbol	Definition
R	charged-particle range; R_M is maximum range; R_{pr} is projected or extrapolated range
RBE	relative biological effectiveness
r	as a subscript, reflected
S	shielding thickness
s	second
s	as a subscript, scattered or skyshine
sr	steradian
T	area-occupancy factor
T	tenth-value layer; T_1 is initial layer; T_e is equilibrium value
$T_{1/2}$	half-life of radioactive decay
t	time
V	volt
v	volume
W	workload factor (for radiotherapy accelerators)
w	aperture width
$X; \dot{X}$	exposure; exposure rate
x	thickness
x	x ray
Y	yield
Z	atomic number
α	reflection coefficient
θ	linear angle
Ω	solid angle
μ	micro- (10^{-6})
γ	gamma ray
Φ	particle fluence
ϕ	particle fluence rate; flux density

APPENDIX B

Data Pertaining to Dose Limits and Radiation Effects on Materials

B.1 Dose-Limiting Recommendations

The indicated values are for the limited scope of this report. NCRP Report No. 39 (NCRP, 1971b) *should* be consulted for more complete information.

Maximum Permissible Dose Equivalent for Occupational Exposure	
Combined whole-body occupational exposure	
Prospective annual limit	5 rems in any one year
Retrospective annual limit	10–15 rems in any one year
Long-term accumulation to age N years	$(N - 18) \times 5$ rems
Skin	15 rems in any one year
Hands	75 rems in any one year
Forearms	30 rems in any one year
Lenses of eyes	5 rems in any one year
Gonads	5 rems in any one year
Red bone marrow	5 rems in any one year
Other systems, tissues and organ systems	15 rems in any one year
Fertile women (with respect to fetus)	0.5 rem in gestation period
Dose Limits for the Public, or Occasionally Exposed Individuals	
Individual or occasional	0.5 rem in any one year
Students	0.1 rem in any one year
Population Dose Limits (averaged over the population)	
Genetic	0.17 rem in a year
Somatic	0.17 rem in a year

B.2 Quality Factors for X Rays and Electrons

X Rays: Q is assumed to be unity. For the purposes of this report:

1 roentgen (unit of exposure) is numerically equal to

1 rad (unit of absorbed dose in tissue) is numerically equal to

1 rem (unit of dose equivalent)

Electrons: Q is assumed to be unity. For the purposes of this report:

1 rad (in tissue) is numerically equal to 1 rem.

Average absorbed dose index rate in the region where an electron beam is stopped in a mass of low-Z material:

$$\dot{D} = \frac{EI}{R_{\text{pr}}} \times 10^8 \text{ rads s}^{-1}.$$

where
E = electron energy (MeV)
I = current density of electrons impinging on surface of material (mA cm^{-2})
R_{pr} = projected electron range (g cm^{-2}).

Empirical range-energy relationships[a]:
  0.01 MeV $< E <$ 2.5 MeV, for low-Z materials:
   R_{pr} = 0.412 E^n
   n = 1.265 − 0.0954 ln E
  2.5 MeV $< E <$ 20 MeV, for low-Z materials:
   R_{pr} = 0.53 E − 0.106

[a] From Katz and Penfold (1952).

B.3 Mean Quality Factors, $\bar{Q}$,[a] and Fluence Rates per Unit Dose-Equivalent Rate for Monoenergetic Neutrons[b]

Neutron energy MeV	$\bar{Q}$	Fluence rate per unit dose-equivalent rate $cm^{-2}\,s^{-1}/mrem\,h^{-1}$
2.5×10^{-8} (thermal)	2	270
1×10^{-7}	2	270
1×10^{-6}	2	220
1×10^{-5}	2	220
1×10^{-4}	2	230
1×10^{-3}	2	270
1×10^{-2}	2.5	280
1×10^{-1}	7.5	46
5×10^{-1}	11	11
1	11	7.6
2.5	9	8.0
5	8	6.4
7	7	6.8
10	6.5	6.8
14	7.5	4.8
20	8	4.4
40	7	4.0
60	5.5	4.4
100	4	5.6

[a] Value of quality factor at the point where the dose equivalent is maximum in a 30-cm tissue-equivalent phantom.

[b] Monoenergetic neutrons incident normally on a 30-cm-thick tissue-equivalent phantom.

Table adapted from ICRU Report No. 20 (ICRU, 1971b).

B.4 Thresholds for Radiation Damage to Selected Materials and Systems[a]

Material	X-Ray and Electron Absorbed Dose
	rads
Transistor	$1 \times 10^3 - 2 \times 10^4$
Potentiometer	$1 \times 10^3 - 1 \times 10^6$
Resistor	$5 \times 10^3 - 1 \times 10^9$
Diode	$1 \times 10^4 - 3 \times 10^4$
Microcircuit	$1 \times 10^4 - 1 \times 10^5$
Glass changes color	$1 \times 10^5 - 4 \times 10^5$
Plastics lose tensile strength	$1 \times 10^6 - 1 \times 10^9$
Natural rubber loses elasticity	$5 \times 10^6 - 3 \times 10^7$
Polymers and oils unusable	$1 \times 10^{10} - 3 \times 10^{10}$
Disinfestation	$10^4 - 10^5$
Pasteurization	$10^5 - 10^6$
Sterilization	$10^6 - 10^7$

	Fast Neutron Fluence
	cm^{-2}
Semiconductor devices	
silicon transistors	$1 \times 10^{12} - 3 \times 10^{14}$
diodes	$1 \times 10^{12} - 3 \times 10^{14}$
tunnel diodes	$1 \times 10^{14} - 1 \times 10^{15}$
germanium transistors	$1 \times 10^{13} - 1 \times 10^{15}$
silicon-carbide diodes	$1 \times 10^{15} - 1 \times 10^{16}$
Carbon potentiometers	$1 \times 10^{15} - 1 \times 10^{16}$
Capacitors	
boron, electrolytic	$1 \times 10^{13} - 1 \times 10^{14}$
aluminum, electrolytic	$1 \times 10^{13} - 1 \times 10^{14}$
tantalum, electrolytic	$1 \times 10^{15} - 1 \times 10^{16}$
paper, oil	$1 \times 10^{15} - 1 \times 10^{16}$
Vacuum tubes	
gas-filled	$1 \times 10^{15} - 1 \times 10^{16}$
power	$1 \times 10^{15} - 1 \times 10^{16}$
photomultiplier	$1 \times 10^{15} - 1 \times 10^{16}$
Insulators, electrical	
Teflon	$1 \times 10^{12} - 1 \times 10^{13}$
Bakelite	$1 \times 10^{13} - 1 \times 10^{14}$
Nylon	$1 \times 10^{14} - 1 \times 10^{15}$

Material	Neutron Fluence
	cm^{-2}
Crystals	
Rochelle salts	$1 \times 10^{12} - 1 \times 10^{13}$
barium titanate	$1 \times 10^{13} - 1 \times 10^{14}$

[a] Bolt and Caroll (1963); DASA (1969); DNA (1972); Kircher and Bowman (1964); Kohl *et al.* (1961).

B.5 Typical Workload, W (for Busy Radiotherapy Installations Only)

Below 10 MeV: refer to Table 2, NCRP Report No. 49 (NCRP, 1976).

Above 10 MeV: 50,000 rem per week at a meter (rem m^2 week^{-1}).

B.6 Area-Occupancy Factor, T^a

Nonoccupationally Exposed Persons (for all accelerators)
   $T = 1$: (full occupancy) work areas such as offices, laboratories, shops, wards, nurses' stations; living quarters; children's play areas; and occupied space in nearby buildings.
   $T = 1/4$: (partial occupancy) corridors, rest rooms, elevators using operators, unattended parking lots.
   $T = 1/16$: (occasional occupancy) waiting rooms, toilets, stairways, unattended elevators, janitors' closets, outside areas used only for pedestrians or vehicular traffic.
   The use of occupancy factors for nonoccupationally exposed persons assumes that only a small portion of the total population is exposed and, hence, the genetically significant dose is small.

Occupationally Exposed Persons
   The occupancy factor of *occupationally* exposed persons, in general, may be assumed to unity.

The above factors are for use as a guide in planning shielding where complete data are not available with respect to accelerator operation, x-ray or electron beam utilization, or area occupancy.

[a] This information is adapted from NCRP Report No. 49 (NCRP, 1976).

APPENDIX C

Conversion Factors and Equivalents

English to Metric System	Metric to English System
Linear Dimensions	
1 inch (in) : 2.54 cm	1 cm : 0.394 inch (in)
1 foot (ft) : 30.5 cm	10 cm : 0.328 foot (ft)
1 yard (yd) : 91.4 cm	100 cm : 1.09 yards (yd)
Weights	
1 ounce (oz): 28.3 g (avoir.)	1 g : 0.0353 ounce (oz) (avoir.)
1 pound (lb): 0.454 kg	1 kg : 2.20 pound (lb)
Mass Thicknesses and Density	
1 lb ft^{-2} : 0.488 g cm^{-2}	1 g cm^{-2}: 2.05 lb ft^{-2}
1 lb in^{-2} : 70.3 g cm^{-2}	1 g cm^{-2} : 0.0142 lb in^{-2}
1 lb ft^{-3} : 0.0160 g cm^{-3}	1 g cm^{-3} : 62.4 lb ft^{-3}

Work Week

1 work week : 40 hours within 7 consecutive days
    2.40×10^3 minutes
    1.44×10^5 seconds

Radiation Beam Equivalents

1 million eV   (MeV): 1.60×10^{-6} erg
1-milliampere beam (mA) : 6.28×10^{15} elementary charges s^{-1}
1-kilowatt beam   (kW) : 1 MeV × mA
        1×10^{10} ergs s^{-1}
        1×10^8 g rads s^{-1}
1 joule   (J) : 1×10^7 ergs
        1 W s
        1×10^5 g rads
1 rad   (rad) : 100 ergs g^{-1}
        1×10^{-5} W s g^{-1}
        1×10^{-2} J kg^{-1}

Attenuation Equivalents

1 tenth-value layer : 3.32 half-value layers
1 half-value layer : 0.301 tenth-value layer

APPENDIX D

Charged Particles

D.1 Range of Monoenergetic Electrons

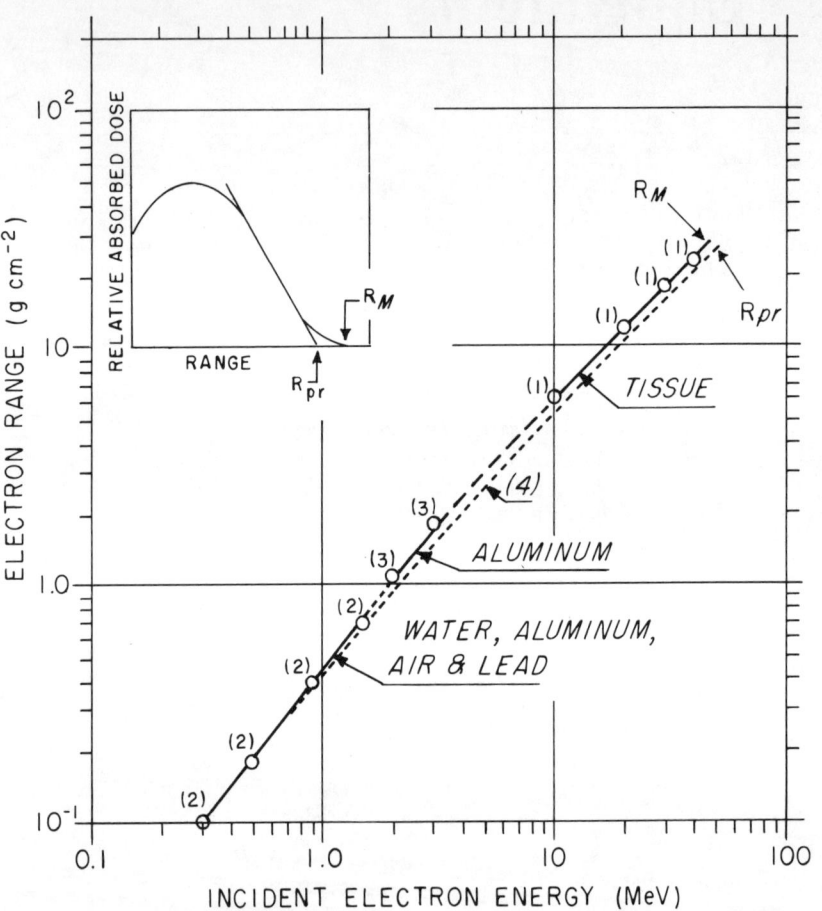

Range of monoenergetic electrons in several materials, as a function of incident electron energy. Data (in units of g cm^{-2}) were derived from curves of absorbed dose versus material thickness given in: (1) Loevinger *et al*. (1961); (2) Trump *et al*. (1940); and (3) Trump *et al*. (1950). The maximum range, R_M, is the intercept of the absorbed-dose/depth curve with the x-ray background, as shown in the insert diagram. For comparison, the projected or extrapolated range, R_{pr}, is plotted as a dotted curve from the empirical formula of Katz and Penfold (1952). Material thickness in cm can be obtained by dividing the range (g cm^{-2}) by the material density (g cm^{-3}). For purposes of shielding calculations, the maximum range *should* be used.

D.2 Fraction of Electrons Backscattered

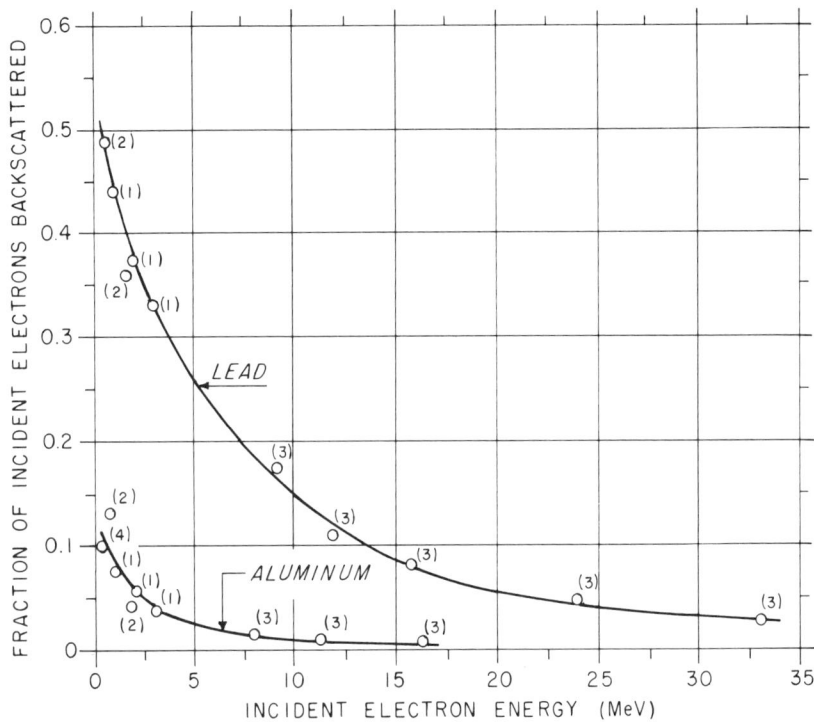

Fraction of incident monoenergetic electrons backscattered from aluminum and lead slabs, as a function of incident electron energy. The included solid angle of backscattering is approximately 2π. Data were obtained from: (1) Wright and Trump (1962); (2) Frank (1959); (3) Harder (1965); and (4) Trump and Van de Graaff (1949). Backscattered electrons retain an appreciable fraction of their incident energy. Typical mean values for the backscattered energy are as follows:

        Pb: 0.75 MeV at 1 MeV;   3.3 MeV at 10 MeV.
        Al: 0.45 MeV at 1 MeV.
        C: ~3.3 MeV at 10 MeV.

Some electrons are backscattered without significant loss in energy. For purposes of shielding calculations, therefore, the incident energy *should* be used, regardless of the number of backscattering processes that are involved.

D.3 Range of Protons

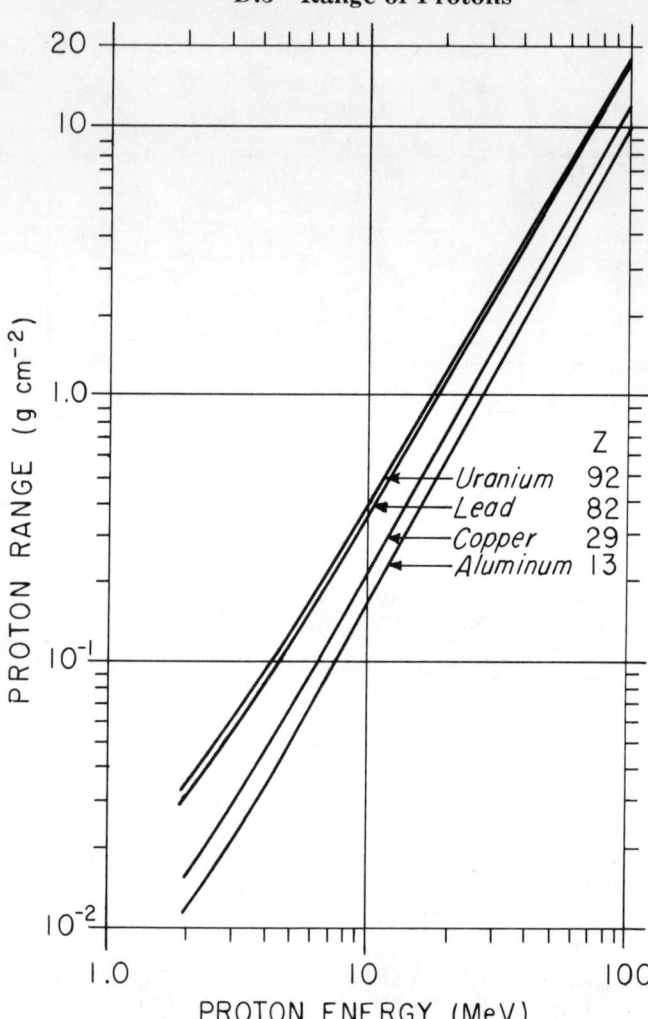

Range of protons (R_p) in several materials, as a function of incident proton energy. Data were obtained from Barkas and Berger (1964). Ranges of other light ions of a given energy, E, can be estimated from the following relationships:

Deuterons	(^{2}H$^+$):	$2\,R_p$ (at $E/2$)
Tritons	(^{3}H$^+$):	$3\,R_p$ (at $E/3$)
Helium-3	(^{3}He$^+$):	$3\,R_p$ (at $E/3$)
	(^{3}He^{++}):	$(3/4)\,R_p$ (at $E/3$)
Helium-4	(^{4}He$^+$):	$4\,R_p$ (at $E/4$)
	(^{4}He^{++}):	R_p (at $E/4$)

These relationships were derived from Spinks and Woods (1964). *Example:* A 4-MeV deuteron has a range of 2 times the range of a 2-MeV proton, i.e. $2 \times 1.2 \times 10^{-2} = 2.4 \times 10^{-2}$ g cm^{-2} in aluminum (density 2.7 g cm^{-3}), or 8.9×10^{-3} cm thickness.

APPENDIX E

X Rays and Gamma Rays

E.1 X-Ray Emission Rates from High-Z Targets

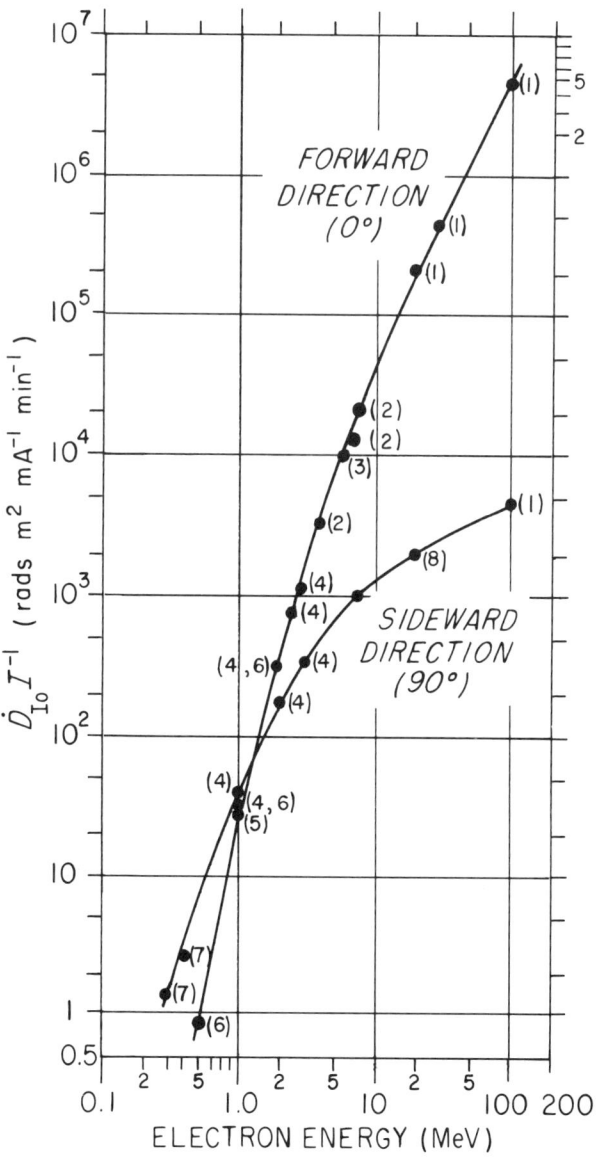

96 / APPENDIX E

X-ray emission rates from normal incidence of monoenergetic electrons on very thick high-Z targets (Z > 73). Emission rates are in terms of absorbed dose index rate, D_{lo}, measured at a standard reference distance of 1 meter, divided by the target current, I (rads m² mA⁻¹ min⁻¹). These data represent the maximum absorbed dose index rates within a 30-cm diameter tissue-equivalent sphere. Selected data points represent emission rates in forward (0°) and sideward (~90°) directions from targets that are optimized in thickness for the highest practical emission rates (e.g., see Berger and Seltzer, 1970). Except for References 1 and 7, these curves are adapted from Bly and Burrill (1959).

(1) Wyckoff *et al*. (1971); (2) Miller (1954); (3) Goldie *et al*. (1954); (4) Burrill (1972); (5) Bly (1964); Cleland (1961); (6) Buechner *et al*. (1948a); (7) McMaster (1963); (8) NCRP Report No. 14 (NCRP, 1954a).

E.2 Angular Distribution of Emitted X Rays from High-Z Targets

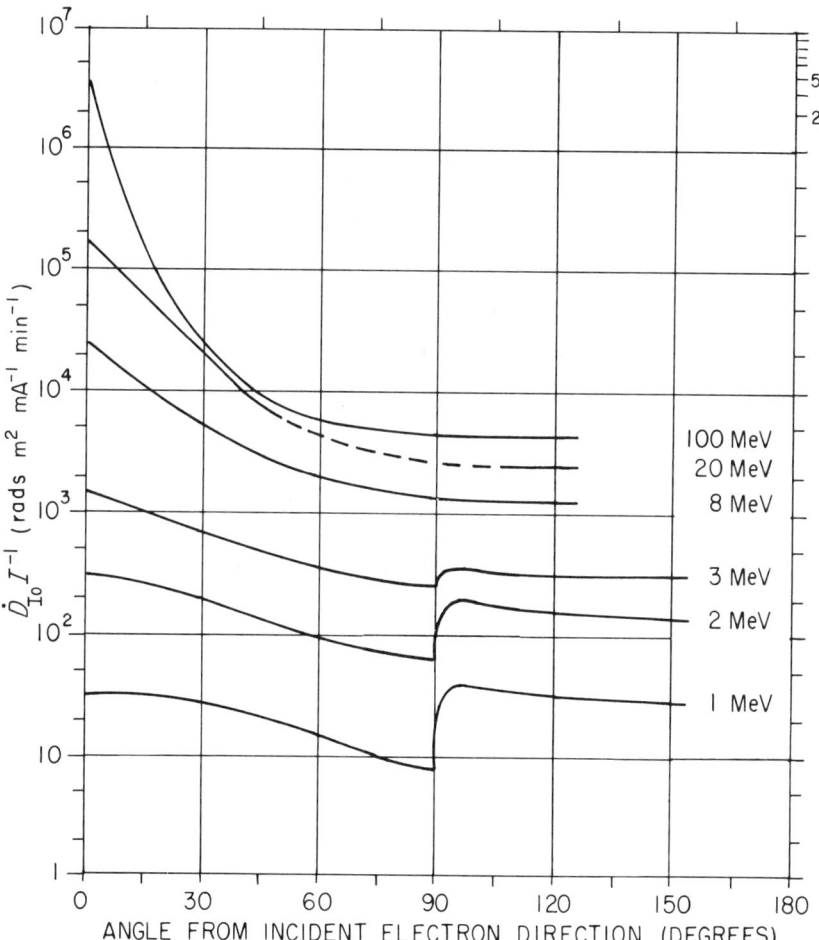

Angular distribution in x-ray emission rate from electrons incident on very thick high-Z targets (see Appendix E-1 caption for explanation of units). The pronounced dip at ~90° in the low-energy curves is due to self-absorption in the target. Because of the extended target geometry for the high-energy measurements, these curves have been smoothed at ~90°. Data were obtained from Bly and Burrill (1959) for the 1- to 8-MeV curves; from Wyckoff et al. (1971) for the 20- and 100-MeV curves.

E.3 X-Ray Emission Rates from Electron Impingement on Thick Targets of Low-Z Material

The x-ray emission rate in the forward (0°) direction is a slowly varying function of Z, e.g., at low electron energies (1.25 to 2.35 MeV), it is approximately proportional to $Z^{1/2}$ (Buechner et al., 1948b). At high energies (up to 100 MeV), this approximation also appears to be reasonable (NCRP Report No. 31, NCRP, 1964b). The 0°-emission rates in Appendix E-1 should be multiplied by the following factors for the indicated target materials:

Target	Z	Factor
Iron or Copper	26 or 29	0.7
Aluminum (concrete)	13	0.5

The x-ray emission rate in the sideward (90°) direction is more Z-dependent, particularly at low electron energies (<10 MeV). For these low energies, the following multiplying factors should be applied to the 90° emission rates in Appendix E-1:

Target	Z	Factor
Iron or copper	26 or 29	0.5
Aluminum (concrete)	13	0.3

At electron energies above 10 MeV, the sideward (90°) emission rates in Appendix E-1 should be used for *all* target materials.

E.4 X-Ray Emission Rates from Electron Backstreaming in Direct Proton Accelerators

Estimated from radiation field maps of 13 Van de Graaff accelerators (Burrill, 1966); measurements consistent within a factor of 2, depending mainly on vacuum conditions and acceleration-tube design; highest values are tabulated below; attenuation by accelerator materials has been included.

Proton energy (MeV)	$\dot{D}_{\infty} I^{-1}$ at angle with respect to directions of proton beams[a] mrads m² μA⁻¹ h⁻¹ [b]	
	90°	180°
0.4	0.5	1
1.0	3	30
2.0	25	200
3.0	50	300
5.0	250	3000

[a] The x-ray source is in the region of the ion source.
[b] See Appendix E-1 caption for explanation of units.

E.5 Radiations from Two-Stage Tandem Accelerators

Estimated from radiation surveys made at several tandem accelerator laboratories, representing extreme conditions (Burrill, 1965); x-ray emission rate may be reduced by improved acceleration-tube design; x-ray emission rate may increase with heavy-ion acceleration due to greater production of secondary electrons within acceleration-tube region; attenuation by accelerator materials has been included.

For the case of 14-MeV protons (7-MV potential at mid-terminal):

X Rays:	90° to beam axis, at mid-terminal:	1×10^4 mrads m^2 μA^{-1} h^{-1}	[a]
Neutrons (fast):	90° to beam axis, at mid-terminal:	2×10^4 m^2 cm^{-2} μA^{-1} s^{-1}	[b]
	analyzer chamber:	5×10^5 m^2 cm^{-2} μA^{-1} s^{-1}	[b]
	Be target in experimental area:	1×10^6 m^2 cm^{-2} μA^{-1} s^{-1}	[b]

Note: The 180° radiations are not as important because of self-absorption by the materials in the accelerator.

[a] See Appendix E-1 for explanation of units.
[b] See Appendix F-1 for explanation of units.

E.6 Equivalent Incident Electron Energies

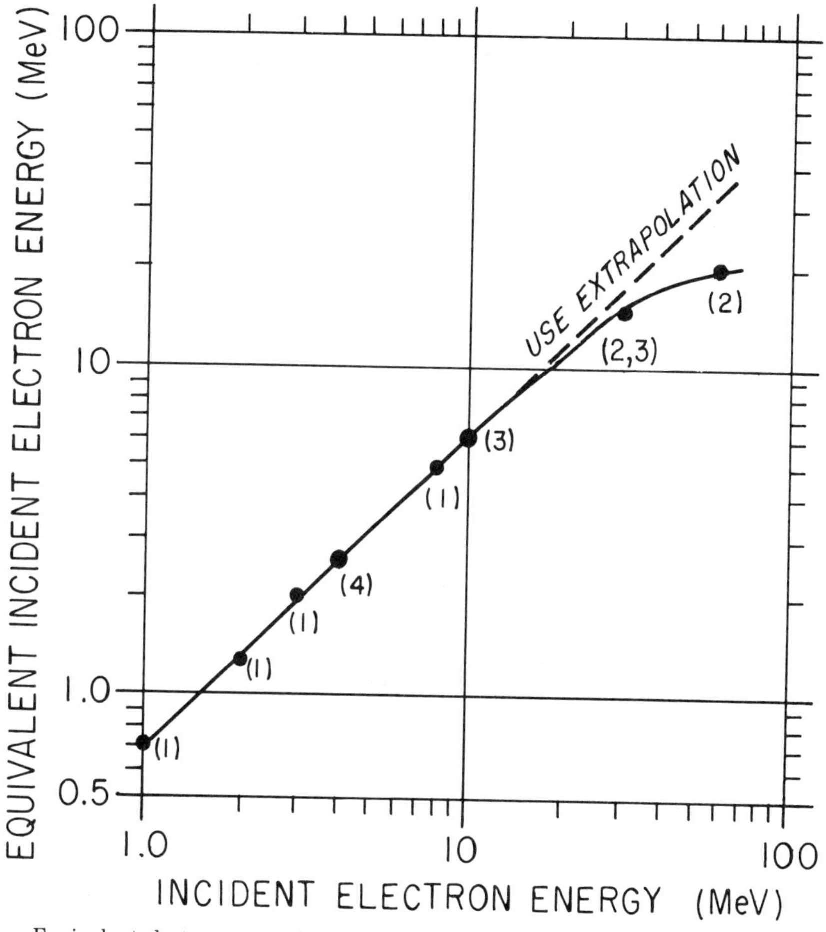

Equivalent electron energy for analysis of transmission of x rays emitted in the 90° direction from very thick high-Z targets, as a function of the incident electron energy. The x-ray spectrum at 90° is lower in energy than the spectrum at 0°. This lower-energy radiation can be described in terms of an incident electron energy that would *in effect* produce x rays with similar transmission characteristics in the 0° direction. Transmission curves or tenth-value layer curves applicable to the lower energy selected from this graph may be used in the calculation of shielding thicknesses for the 90° beam. The same procedure would be a conservative approach for x rays from low-Z targets, and for x rays emitted in the 180° direction.

References: (1) Burrill (1968); (2) Berger and Seltzer (1970); (3) McCall and Nelson (1974); and (4) Saxon (1964).

E.7 Broad-Beam Transmission Through Concrete of X Rays Produced by 0.1- to 0.4-MeV Electrons

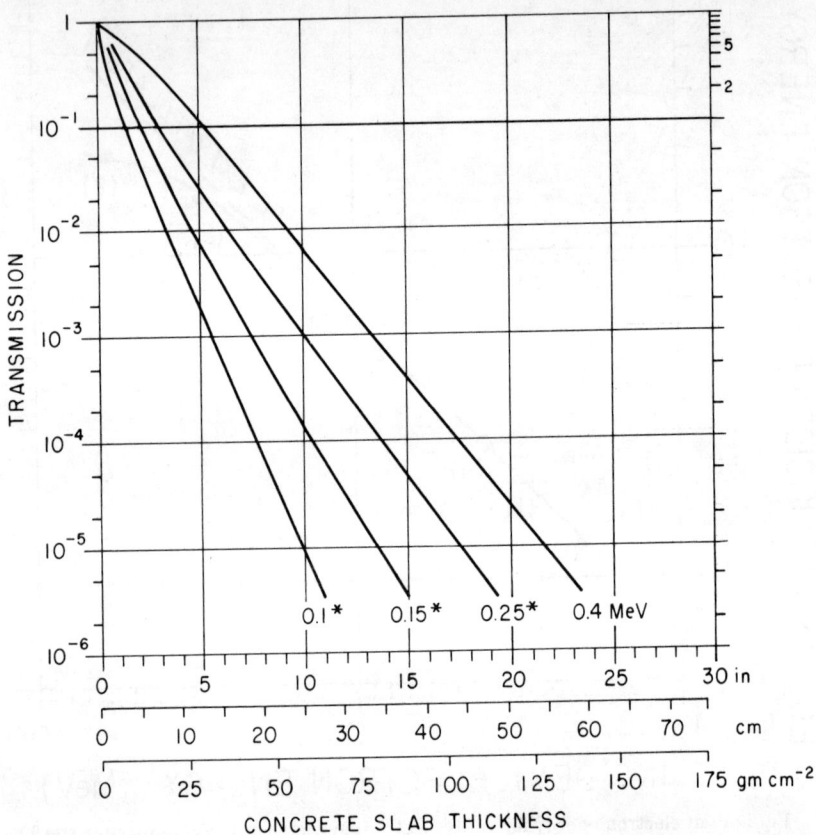

Transmission through concrete (density 2.35 g cm^{-3}) of x rays produced by 0.1- to 0.4-MeV electrons, under broad-beam conditions. Electron energies designated by an asterisk (*) were accelerated by voltages with pulsed wave form; unmarked electron energies were accelerated by a constant potential generator. Curves represent transmission in dose-equivalent index ratio. (See Appendix E-12 for basis of interpolating between curves.) Curves were derived from NCRP Report No. 34 (NCRP, 1970a).

E.8 Broad-Beam Transmission Through Concrete of X Rays Produced by 0.5- to 176-MeV Electrons

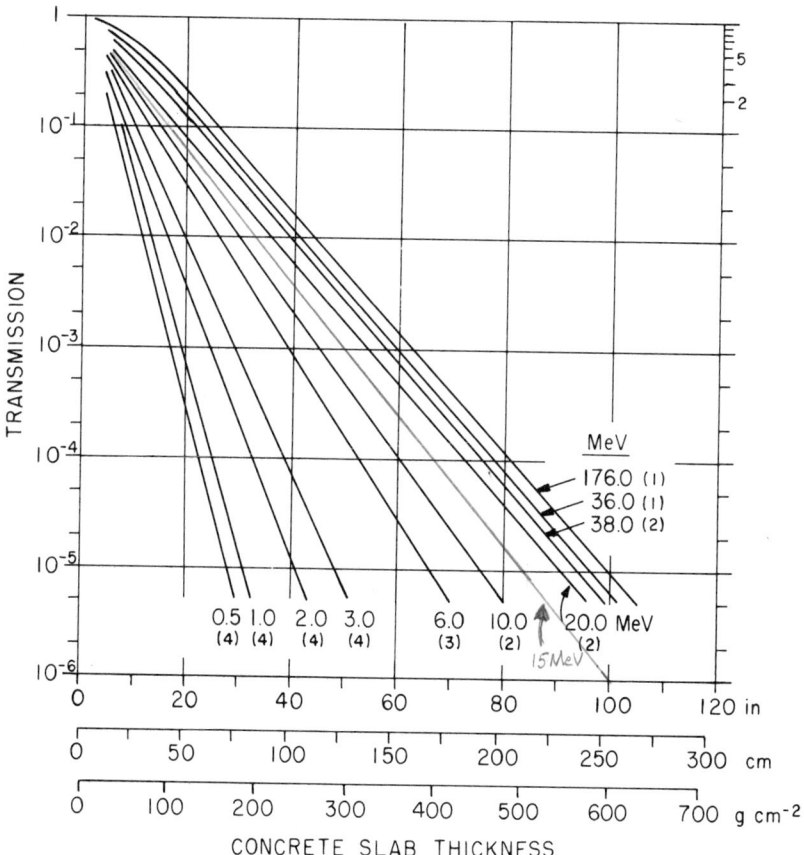

Transmission of thick-target x rays through ordinary concrete (density 2.35 g cm^{-3}), under broad-beam conditions. Energy designations on each curve (0.5 to 176 MeV) refer to the monoenergetic electron energy incident on the thick x-ray producing target. Curves represent transmission in dose-equivalent index ratio. (See Appendix E-12 for basis for interpolating between curves.) Curves derived from: (1) Miller and Kennedy (1956); (2) Kirn and Kennedy (1954); (3) Karzmark and Capone (1968); and (4) NCRP Report No. 34 (NCRP, 1970a) and NCRP Report No. 49 (NCRP, 1976).

E.9 Broad-Beam Transmission Through Steel of X Rays Produced by 1- to 31-MeV Electrons

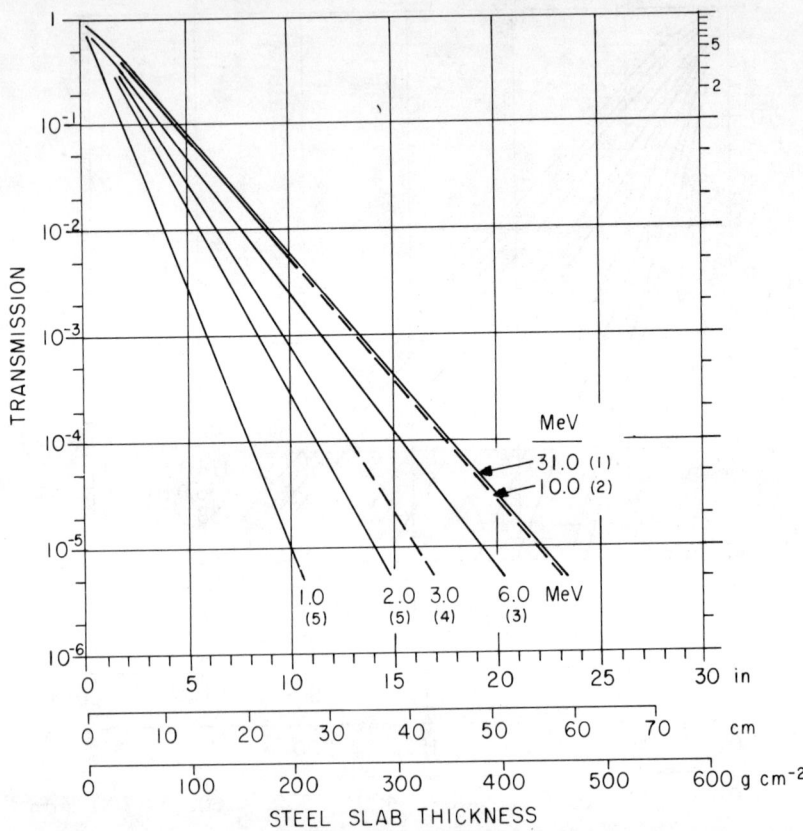

Transmission of thick-target x rays through steel (density 7.8 g cm^{-3}), under broad-beam conditions. Energy designations on each curve refer to the monoenergetic electron energy incident on the thick x-ray producing target. Curves represent transmission in dose-equivalent index ratio. (See Appendix E-13 for basis for interpolating between curves.) Curves were derived from: (1) Wideroë (1953); (2) O'Connor et al. (1949); (3) Karzmark and Capone (1968); (4) Goldie et al. (1954); and (5) Buechner et al. (1948a).

E.10 Broad-Beam Transmission Through Lead of X Rays Produced by 0.1- to 0.4-MeV Electrons

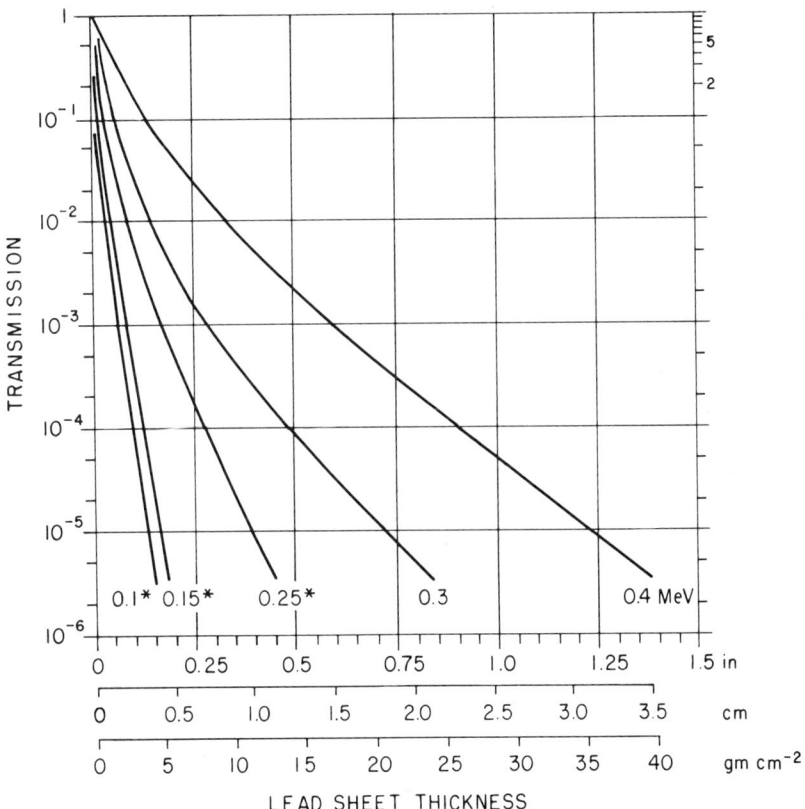

Transmission through lead (density 11.3 g cm^{-3}) of x rays produced by 0.1- to 0.4-MeV electrons, under broad-beam conditions. Electron energies designated by an asterisk (*) were accelerated by voltages with pulsed wave form; unmarked electron energies were accelerated by a constant potential generator. Curves represent transmission in dose-equivalent index ratio. (See Appendix E-14 for basis for interpolating between curves.) Curves were derived from NCRP Report No. 34 (NCRP, 1970a).

E.11 Broad-Beam Transmission Through Lead of X Rays Produced by 0.5- to 86-MeV Electrons

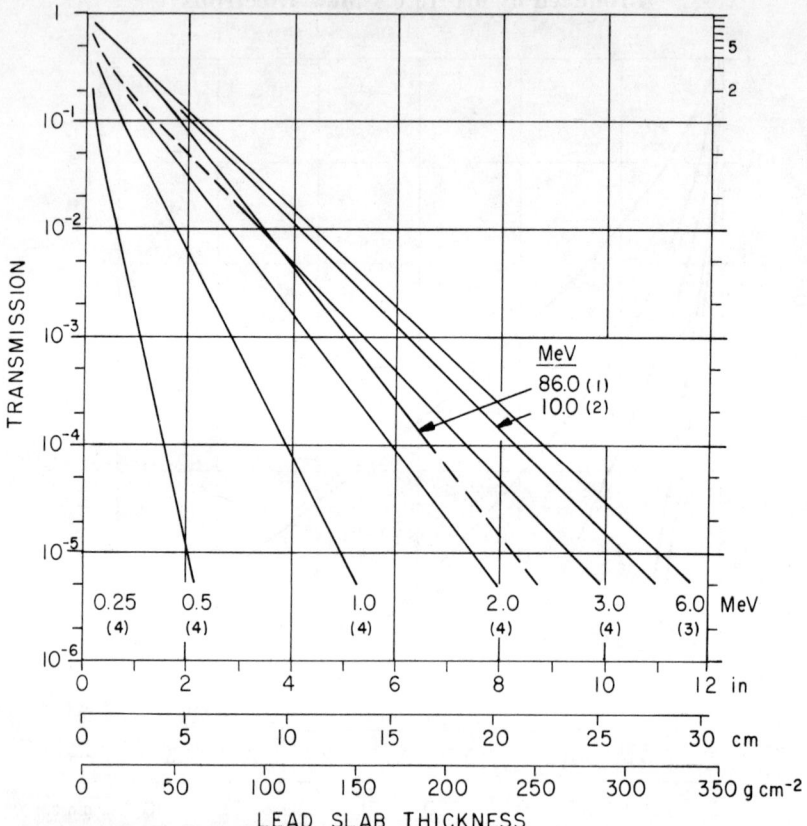

Transmission of thick-target x rays through lead (density 11.3 g cm^{-3}), under broad-beam conditions. Energy designations on each curve (0.5 to 86 MeV) refer to the monoenergetic electron energy incident on the thick x-ray producing target. Curves represent transmission in dose-equivalent index ratio. (See Appendix E-14 for basis for interpolating between curves.) Curves were derived from: (1) Miller and Kennedy (1956); (2) ICRP Publication No. 4 (ICRP, 1964); (3) Karzmark and Capone (1968); and (4) NCRP Report No. 34 (NCRP, 1970a) and NCRP Report No. 49 (NCRP, 1976).

E.12 Dose-Equivalent Index Tenth-Value Layers for Broad-Beam X Rays in Concrete

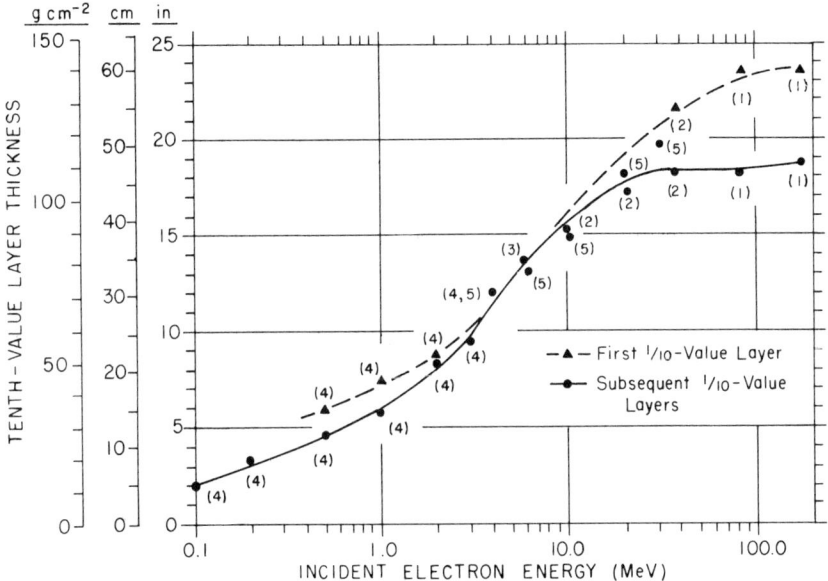

Dose-equivalent index tenth-value layers in ordinary concrete (density 2.35 g cm^{-3}) for thick-target x rays under broad-beam conditions, as a function of the energy of electrons incident on the thick target. The dotted curve refers to the first tenth-value layer; the solid curve refers to subsequent or "equilibrium" tenth-value layers. Both curves are empirically drawn through data points derived from the following references: (1) Miller and Kennedy (1956); (2) Kirn and Kennedy (1954); (3) Karzmark and Capone (1968); (4) NCRP Report No. 34 (NCRP, 1970a); (5) Maruyama et al. (1971). Studies by Lokan et al. (1972) on light Ilmenite-loaded concrete (density 2.89 g cm^{-3}) are in reasonable agreement with the solid curve above, on a mass thickness basis (g cm^{-2}).

E.13 Dose-Equivalent Index Tenth-Value Layers for Broad-Beam X Rays in Steel

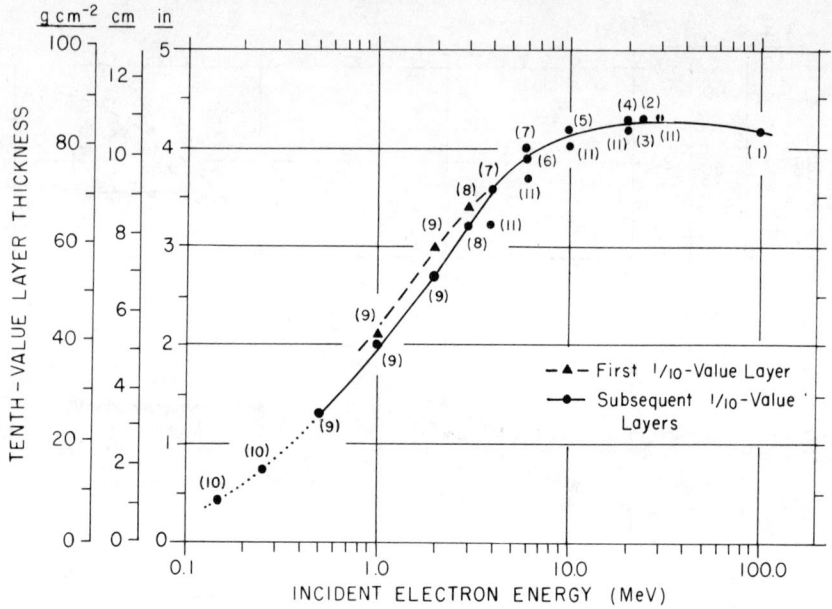

Dose-equivalent index tenth-value layers in steel (density 7.8 g cm^{-3}) for thick-target x rays under broad-beam conditions, as a function of the energy of electrons incident on the thick target. The dotted curve refers to the first tenth-value layer; the solid curve refers to subsequent or "equilibrium" tenth-value layers. Both curves are empirically drawn through data points derived from the following references: (1) Westendorp and Charlton (1945); (2) Wideroë (1953); (3) Scag (1954); (4) Adams and Girard (1946); (5) O'Connor et al. (1949); (6) Karzmark and Capone (1968); (7) NCRP Report No. 34 (NCRP, 1970a); (8) Goldie et al. (1954); (9) Buechner et al. (1948a); (10) estimated from Bly and Burrill (1959); and (11) Maruyama et al. (1971). Studies by Lokan et al. (1972) on heavy Ilmenite-loaded concrete (density 4.30) are in reasonable agreement with the solid curve above, on a mass thickness basis (g cm^{-2}).

E.14 Dose-Equivalent Index Tenth-Value Layers for Broad-Beam X Rays in Lead

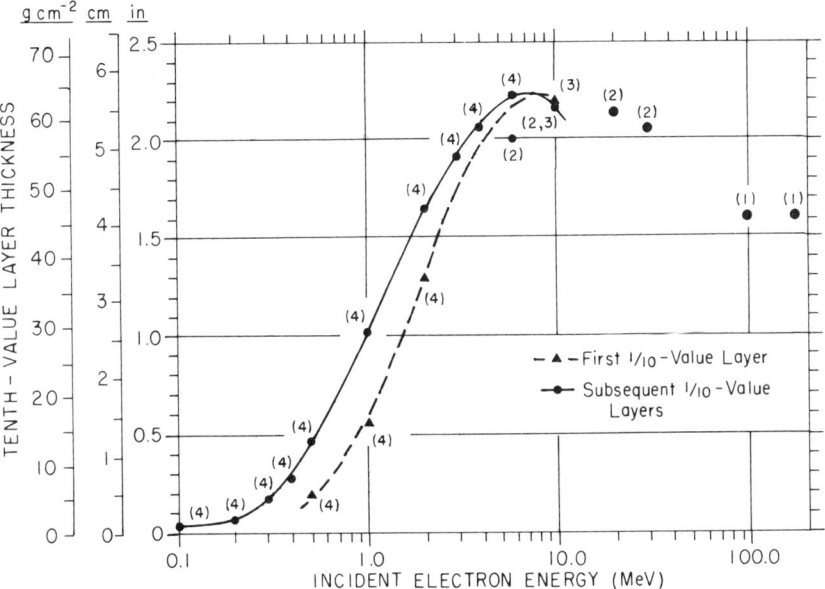

Dose-equivalent index tenth-value layers in lead (density 11.3 g cm^{-3}) for thick-target x rays under broad-beam conditions, as a function of the energy of electrons incident on the thick target. The dotted curve refers to the first tenth-value layer; the solid curve refers to subsequent or "equilibrium" tenth-value layers. Both curves are empirically drawn through data points derived from the following references: (1) Miller and Kennedy (1956); (2) Maruyama *et al.* (1971); (3) ICRP Publication No. 4 (ICRP, 1964); and (4) NCRP Report No. 34 (NCRP, 1970a). The empirical curve is not extended into the 10- to 100-MeV region because of uncertainties in the available data.

E.15 Reflection Coefficients for Monoenergetic X Rays in Concrete, Iron and Lead

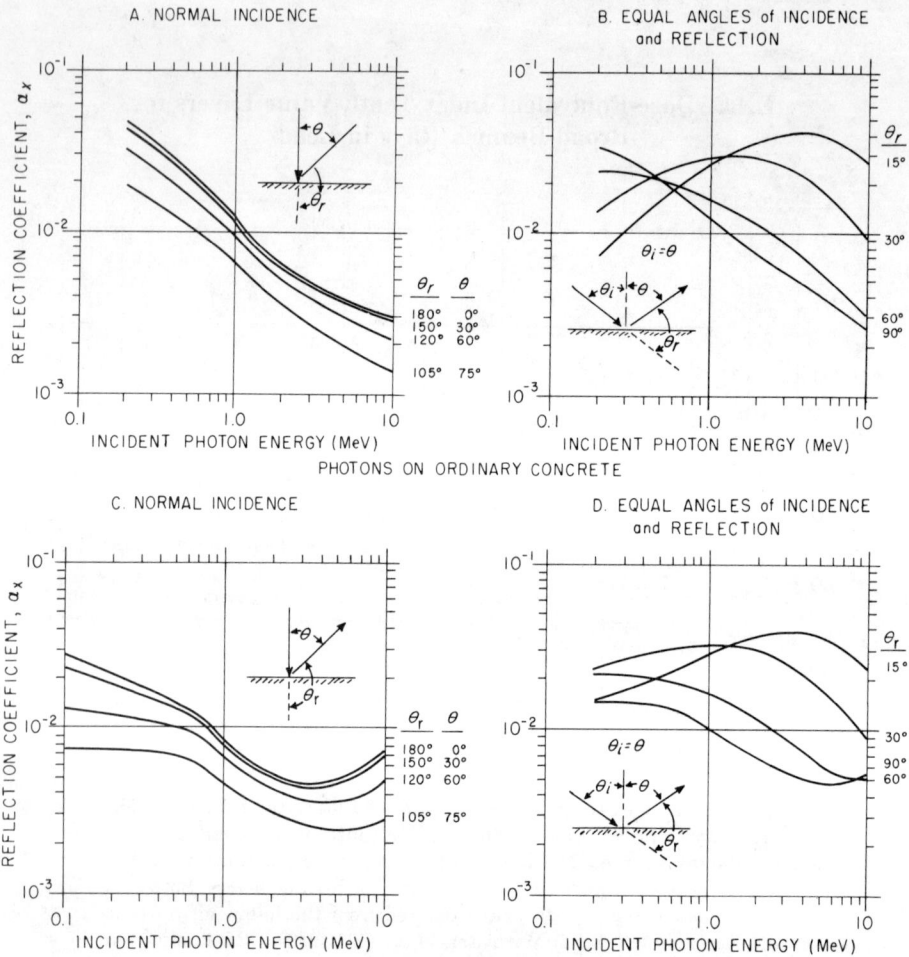

Reflection coefficients, α_x, for monoenergetic x rays on ordinary concrete, iron, and lead as a function of incident monoenergetic photon energy, for several angles of reflection assuming normal incidence and equal angles of incidence and reflection. Values are given for ordinary concrete and iron, based on existing available information, both theoretical and experimental, with particular emphasis on the following references: (1) Chilton and Huddleston (1963); (2) Chilton (1964); (3) Chilton (1965); and (4) Chilton et al. (1965). For photon energies higher than 10 MeV, the use of the 10-MeV values of α_x is expected to be safe.

Values of α_x for photons incident on lead are not as readily calculable, but a conservative upper limit is 5×10^{-3} for any energy and scattering angle.

The values of α_x for $\theta_r = 180°$ in Curve A are the same as for $\theta_r = 180°$ in Curve B.

APPENDIX F

Neutrons

F.1 Thick-Target Neutron Fluence Rates for (p,n) Reactions

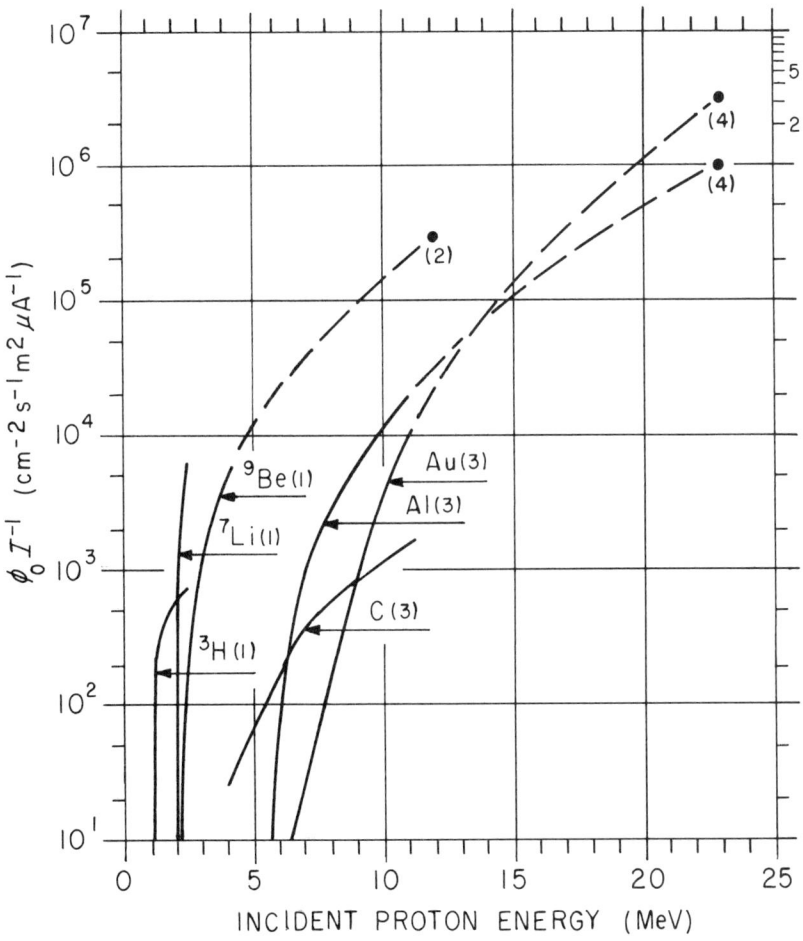

Thick-target neutron fluence rates[a] from several (p,n) reactions, as a function of incident proton energy; measured in the 0° direction at 1 meter from the target and divided by the target current (cm^{-2} s^{-1} m^2 μA^{-1}). The curves were empirically derived from the following references: (1) Burrill (1963); (2) Gove (1956); (3) Bromley et al. (1959); and (4) Fowler and Brolley (1956). Neutron fluence rates at higher incident proton energies are estimated from Amos et al. (1970), as follows:

Proton energy MeV	$\phi_o I^{-1}$ (0°) (cm^{-2} s^{-1} m^2 μA^{-1})		
	C	Al	Ta or Pb
30	1.3×10^6	6×10^6	6×10^6
40	6×10^6	1.3×10^7	2.2×10^7

See Appendix F-4 for estimating neutron fluence rates in other directions from the target.

[a] Fluence rates are derived from yield per steradian (Y sr^{-1}).

F.2 Thick-Target Neutron Yields for (d,n) Reactions

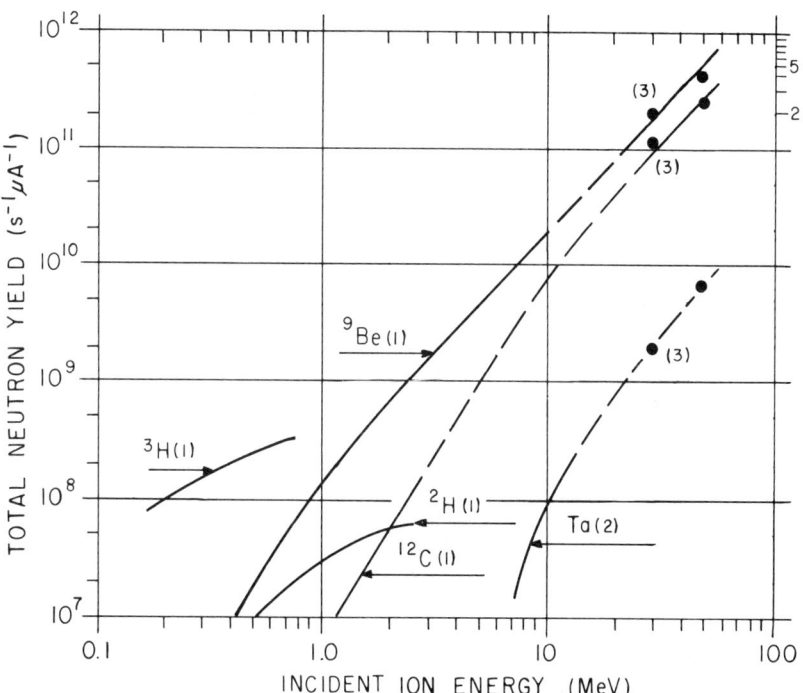

Thick-target neutron yields (in 4π solid angle) per microampere of target current from several (d,n) reactions, as a function of incident deuteron energy. The curves were empirically derived from the following references: (1) Burrill (1963); (2) Borchers and Wood (1965); and (3) Schweimer (1967). Neutron fluence rates can be estimated, for purposes of calculating shielding thicknesses, from Appendix F-4. The yield from the ^{27}Al(d,n) reaction is slightly less than that from the ^{12}C(d,n) reaction.

F.3 Thick-Target Neutron Yields for (γ,n) Reactions

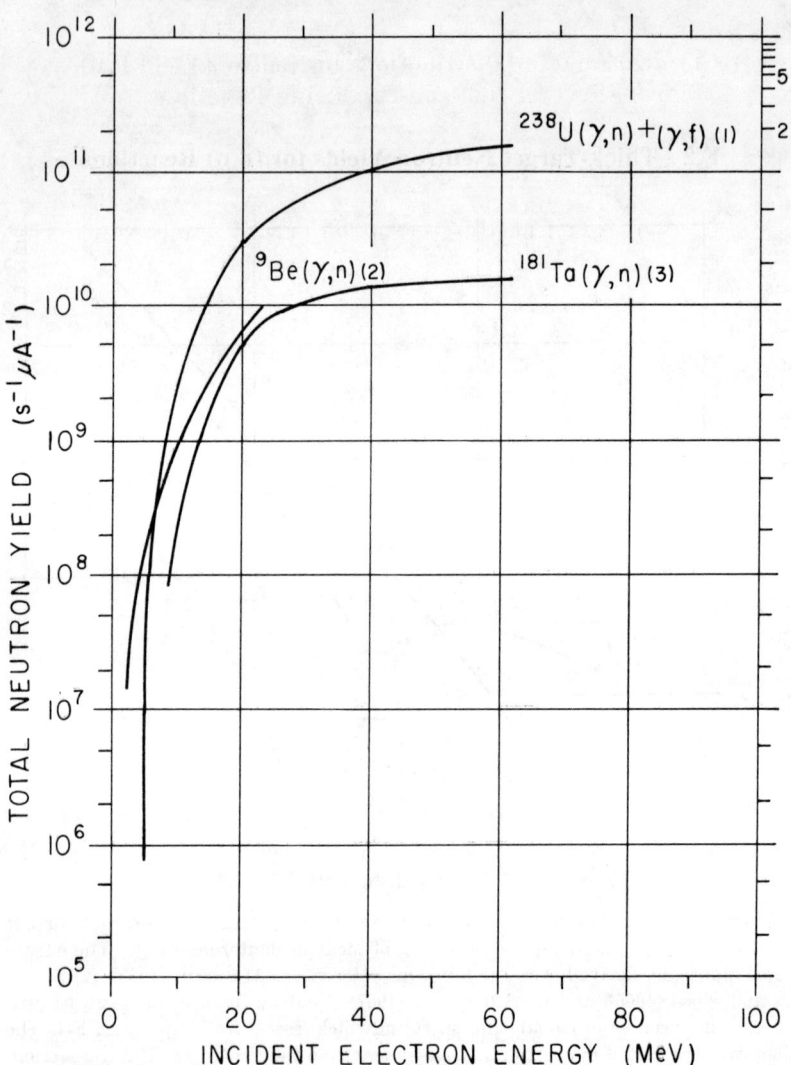

Thick-target neutron yields (in 4π solid angle) per microampere of target current from several photonuclear reactions, as a function of incident electron energy on the thick-target x-ray source. In the case of the ^{9}Be(γ,n) yields, a mass of Be was placed around the x-ray source. The ^{181}Ta target thickness was 6.3 g cm^{-2}, serving as x-ray source as well as neutron source. The ^{238}U target was 5.9 g cm^{-2} thick; the yields include the neutrons from the photofission of ^{238}U. The curves were empirically derived from the following references: (1) Gozani and Kull (1969); (2) MacGregor (1959); and (3) Berger and Seltzer (1970). Neutron fluence rates can be estimated, for purposes of calculating shielding thickness, from Appendix F-4.

F.4 Typical Angular Distributions in Neutron Yield Ratios[e] from Several Neutron-Producing Reactions

Reaction	Energy MeV	A $Y\ sr^{-1}\ (90°)/Y\ sr^{-1}\ (0°)$[b]	B $Y\ sr^{-1}\ (0°)/Y\ sr^{-1}$ [b,c]	Ref.
^{3}H(p,n)^{3}He	≥1.02 (E_{th})[a]	0.25	0.25	1
	2.5	0.5	0.15	1
^{7}Li(p,n)^{7}Be	≥1.88 (E_{th})	0.25	0.25	1
	2.5	0.5	0.15	1
^{9}Be(p,n)^{9}B	≥2.05 (E_{th})	0.25	0.25	1
	2.5	0.67	0.10	1
	14	<0.1	>0.5	2
^{2}H(d,n)^{3}He	0.5	0.3	0.2	1
	2.5	0.15	0.35	1
^{3}H(d,n)^{4}He	0.2	~1	~0.1	1
^{9}Be(d,n)^{10}B	<2	~1	~0.1	1
	7.5	0.1	0.5	2
^{9}Be(^{3}He,n)^{11}C	20	0.1	0.5	2
Ta(^{4}He,n)	40	<0.1	>0.5	3
All (p,n), (d,n) (^{3}He,n), (^{4}He,n)	>5	≤0.1	≥0.5	—
All (γ,n)	≫E_{th}	~2	~0.5[d]	—

[a] E_{th}: threshold energy.

[b] Ratios of yields per steradian are estimates only, and are useful mainly for purposes of calculating shielding thicknesses.

[c] Yield per steradian in all other directions is assumed to be equal to $Y\ sr^{-1}\ (90°)$.

[d] $Y\ sr^{-1}\ (90°)/Y$.

[e] Neutron fluence rates, ϕ, at distance d from the target can be approximated by $(Y\ sr^{-1} \times 10^{-4})/d^2$, when d is measured in meters. Where total (4π) yield is given (Appendices F-2, F-3), $Y\ sr^{-1}\ (0°)$ is obtained from Column B. $Y\ sr^{-1}\ (90°)$ is then obtainable from Column A. Note: for (γ,n) reactions, Column B provides the value for $Y\ sr^{-1}\ (90°)$ directly.

References: (1) Burrill (1963); (2) Bruninx and Crombeen (1969); and (3) Stephens and Miller (1969).

F.5 Maximum Energies of Neutrons Produced from Proton and Deuteron Reactions

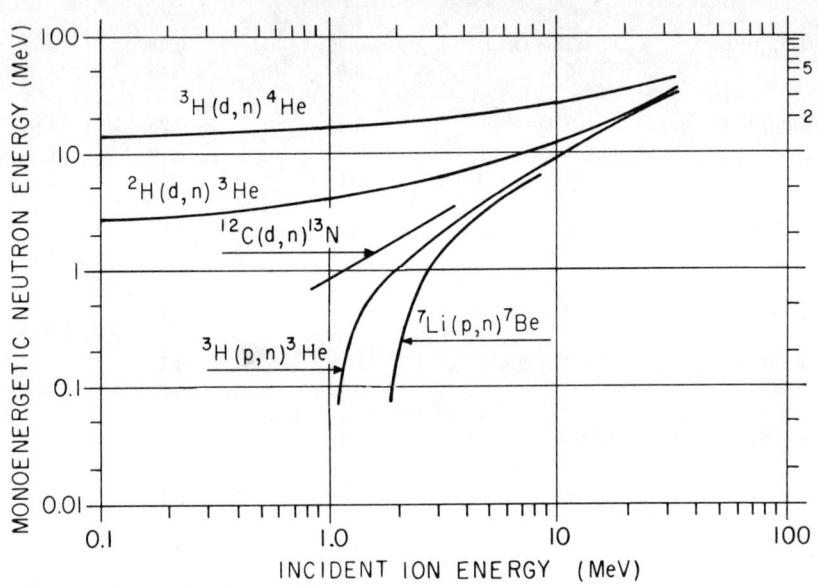

Energies of monoenergetic neutrons (in the laboratory system) from several nuclear reactions, as a function of incident proton or deuteron energy, emitted from the target in the 0° direction. Neutrons emitted in other directions are less energetic. Energy values were obtained from tables in Fowler and Brolley (1956) and Marion and Fowler (1960), except those from the $^{12}C(d,n)^{13}N$ reaction, which were estimated from Hanson and Taschek (1950).

F.6 Dose-Equivalent Index Transmission Through Concrete of Monoenergetic Neutrons

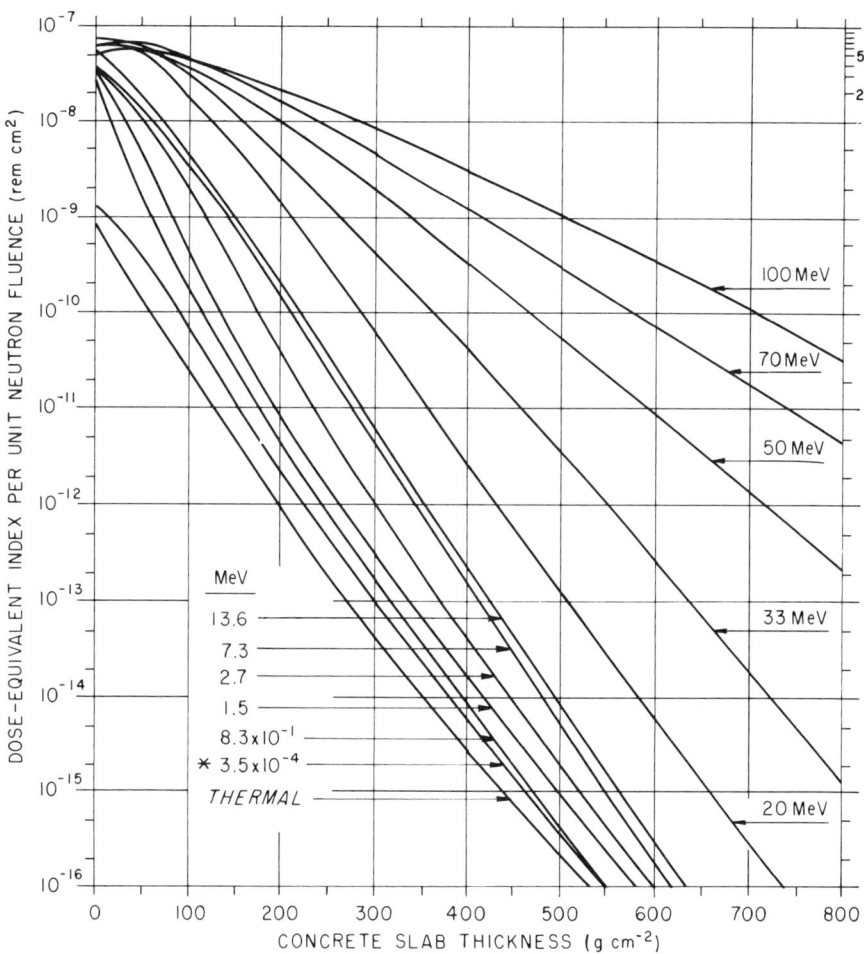

Dose-equivalent index transmission per unit fluence (rem cm²) of monoenergetic neutrons incident normally on slabs of ordinary concrete (TSF-5.5, see Appendix H-2). Multi-collision dose-equivalent index per unit fluence is calculated to include the dose-equivalent index contribution by capture gamma rays produced within the slab. Energies noted on the curves refer to the average energies of neutrons in the energy bins chosen for the discrete-ordinate calculations, as follows:

F.6 Continued

Energy Bin MeV	Average Energy MeV
0–4.14×10^{-7}	"thermal"
1.01×10^{-4}–5.38×10^{-4}	3.5×10^{-4}*
5.50×10^{-1}–1.11	8.3×10^{-1}
1.11–1.83	1.5
2.46–3.01	2.7
6.36–8.19	7.3
12.2–15.0	13.6
15.0–25.0	20
25.0–40.0	33
40.0–60.0	50
60.0–80.0	70
80.0–125.0	100

* The transmission curve for this average energy is typical of transmission curves in the energy range 1×10^{-4} to 5×10^{-1}.

These calculations were prepared by Wyckoff and Chilton (1973), based on Roussin and Schmidt (1971) and Roussin et al. (1973).

F.7 Accelerator Conditions for Generating Neutron Spectra Referred to in Appendices F-8 and F-9

Reaction class	Incident particle energy MeV	Target	Angle to beam degrees	$T_e{}^a$ g cm^{-2}	Reference
(γ,n)	16	Pt	90	78	3
	34	O	55	71	2
	55	Be	67.5	85	4
	55	Pb	67.5	86	4
	65	O	90	84	1
	65	Mg	90	86	1
	85	Be	67.5	163	4
	85	Pb	67.5	117	4
(p,n)	8	C	0	74	6
	8	Ta	0	81	6
	10	Mg	0	78	6
	10	S	0	78	6
	10	Si	0	78	6
	12	Mg	0	78	6
	12	S	0	78	6
	12	Si	0	78	6
	13	C	0	74	6
	13	Ta	0	81	6
	14	Al	10	71	5
	20	Al	10	71	5
(d,n)	1.6	Be	0	74	11
	8	Ta	10	74	8
	14	Ta	10	74	8
	16	Be	0	74 (est.)	10
	18	Al	0	71	5
	40	Be	0	86	9
	54	Be	0	115	9
(^{3}He,n)	18	Cu	—	74	7
(γ,n) + (γ,f)	45	^{238}U	90	85	6

[a] T_e is equilibrium tenth-value layer for concrete as obtained from Appendices F-8 and F-9 (see Section 4.3.3 for limitations in use; see Appendix F-10 for values of T_e for monoenergetic neutrons).

References: (1) Firk (1967); (2) Verbinski and Courtney (1965); (3) Glazunov et al. (1964); (4) Kaushal et al. (1968); (5) Holeman et al. (1969); (6) ICRU (1969); (7) McCaslin and Smith (1969); (8) Borchers and Wood (1965); (9) Schweimer (1967); (10) Fleischer (1968); and (11) Shpetnyi (1957).

F.8 Dose-Equivalent Index Transmission Through Concrete of Neutrons from (γ,n) and (γ,fn) Reactions

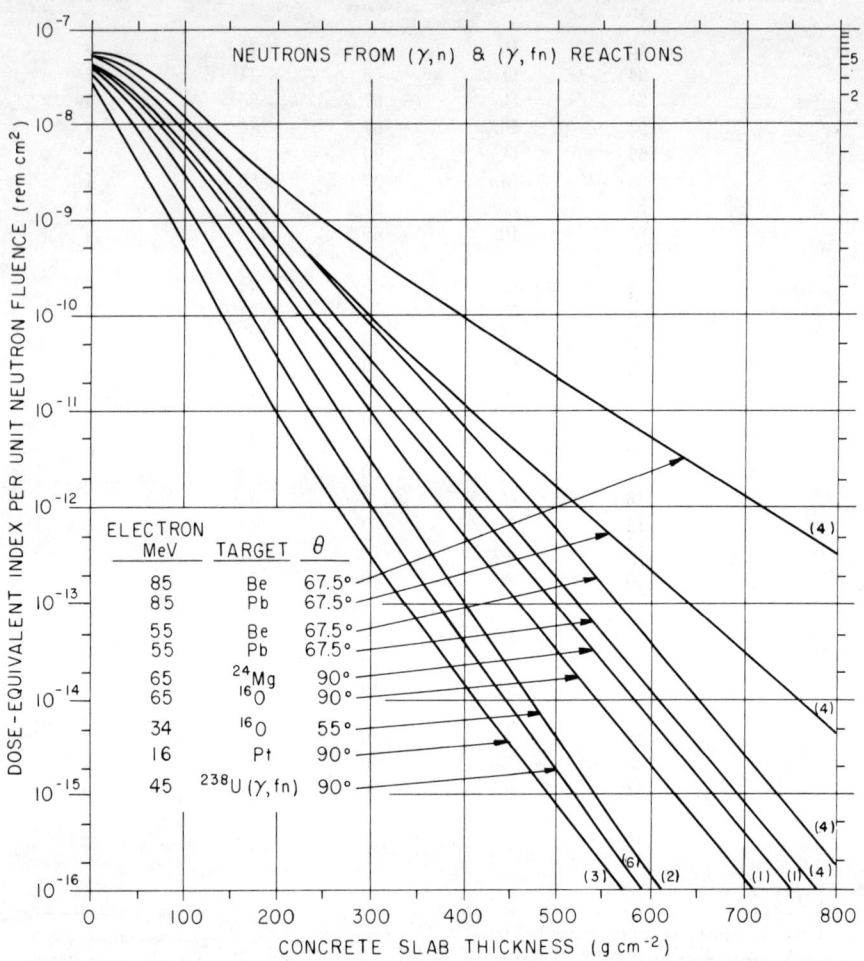

Dose-equivalent index transmission per unit fluence (rem cm^2) of neutrons from (γ,n) and (γ,fn) reactions, incident normally on slabs of ordinary concrete (TSF-5.5, see Appendix H-2). Multi-collision dose-equivalent index was derived by folding individual incident neutron spectra (Appendix F-7) into monoenergetic-neutron transmission data (Appendix F-6). (See Appendix F-7 for references.)

F.9 Dose-Equivalent Index Transmission Through Concrete of Neutrons from Ion-Induced Reactions

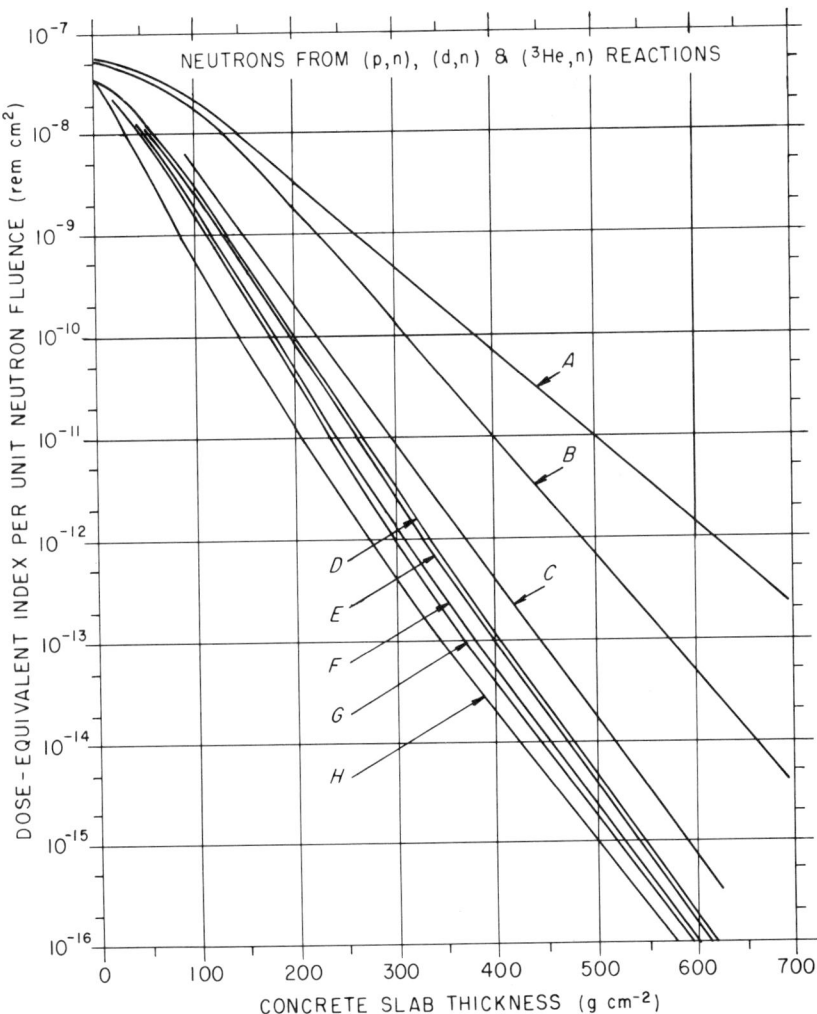

Dose-equivalent index transmission per unit fluence (rem cm^2) of neutrons from several ion-induced reactions, incident normally on slabs of ordinary concrete (TSF-5.5, see Appendix H-2). Multi-collision dose-equivalent index was derived by folding individual incident neutron spectra (Appendix F-7) into monoenergetic-neutron transmission data (Appendix F-6). The table below presents the reactions from which the curves were constructed.

F.9 Continued

Reaction class	Ion energy MeV	Target	Curve	Reference[a]
(d,n)	54	Be	A	9
	40	Be	B	9
	16	Be	C (est.)	10
	1.6	Be	F	11
	18	Al	D	5
	14	Ta	F	8
	8	Ta	G	8
(p,n)	20	Al	D	5
	14	Al	E	5
	13	C	F	6
	8	C	G	6
	12, 10	Mg, S, Si	G-H[b]	6
(^{3}He,n)	18	Cu	F	7

[a] See Appendix F-7 for references.
[b] These data lie on curves intermediate between G and H.

F.10 Dose-Equivalent Index Tenth-Value Layers in Concrete for Monoenergetic Neutrons

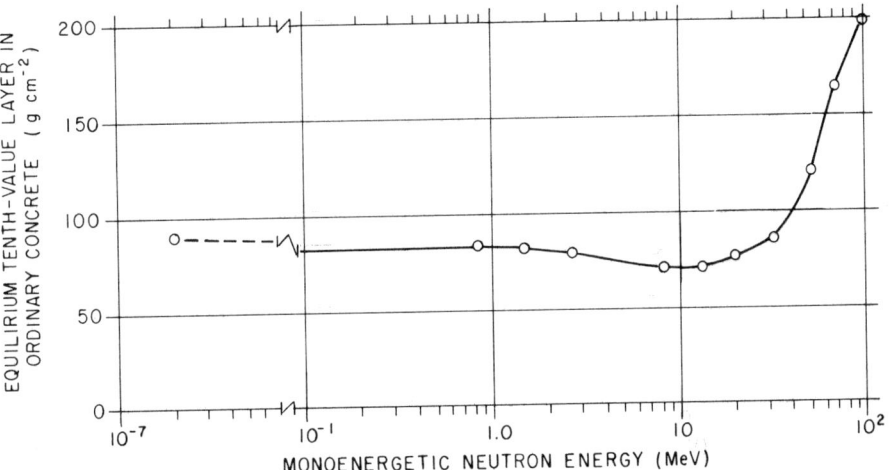

Equilibrium dose-equivalent index tenth-value layers, T_e, for neutrons in ordinary concrete, as a function of monoenergetic-neutron energy. T_e is applicable only for transmission *less* than 10^{-16} rem cm². The curve above is broken between 10^{-6} and 10^{-1} MeV to indicate that the trend in T_e is very slow moving in this energy range. Data were obtained directly from curves in Appendix F-6.

The equilibrium dose-equivalent index tenth-value layers for Appendices F-8 and F-9 are listed in Appendix F-7, accompanying the parameters of the corresponding neutron spectrum. Except for (γ,n) spectra produced by electron energies greater than 55 MeV, and for neutrons produced by charged-particle energies greater than 20 MeV, the values of T_e lie below 85 g cm^{-2}.

Note: Refer to Section 4.3.3 regarding the limited usefulness of these data.

F.11 Thermal-Neutron Transmission Through Mazes and Ducts

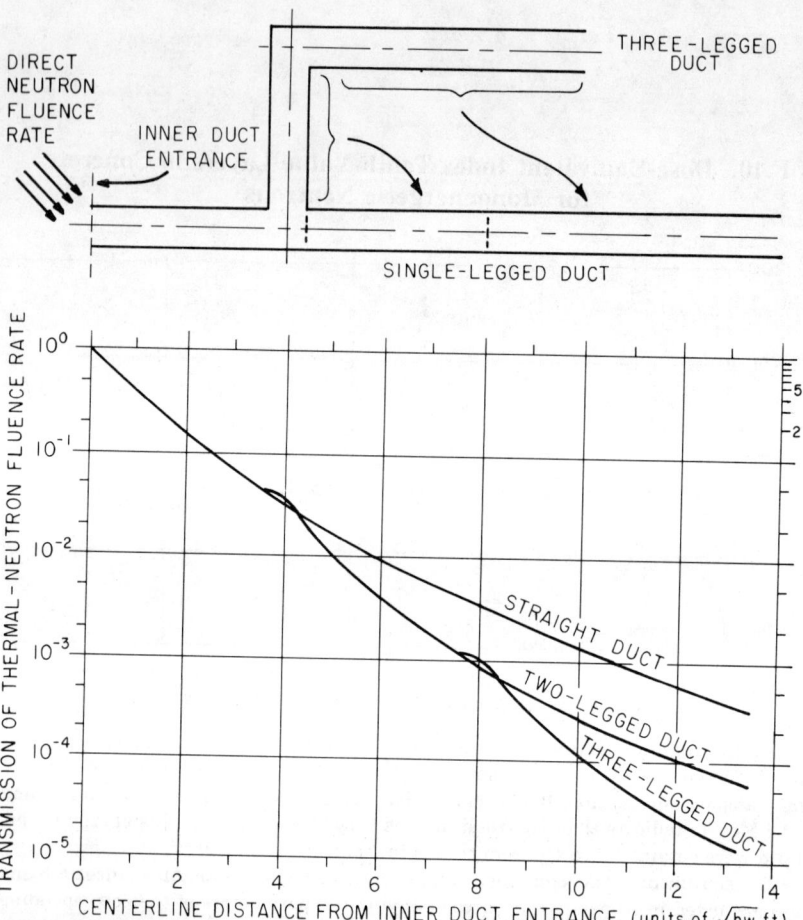

Transmission of thermal-neutron fluence rates through multi-legged concrete mazes or large ducts, as a function of the summed centerline distances from the inner duct entrance to the duct exit, divided by the square root of the product of the height and width of the duct aperture ($\sqrt{hw}$ ft). The curves for each leg are essentially similar in shape, with discontinuities in the transmission factor at the intersections between legs. The curve for the second leg intersects the first-leg curve at a distance equal to the centerline length of the first leg; the curve for the third leg intersects similarly on the second-leg curve. These curves are adapted from Maerker *et al.* (1968), and are illustrative of a specific maze configuration. Intersects of curves for other configurations would occur at dimensions corresponding to the particular maze design in question.

F.12 Reflection Coefficients for Monoenergetic Neutrons in Concrete, Iron and Lead

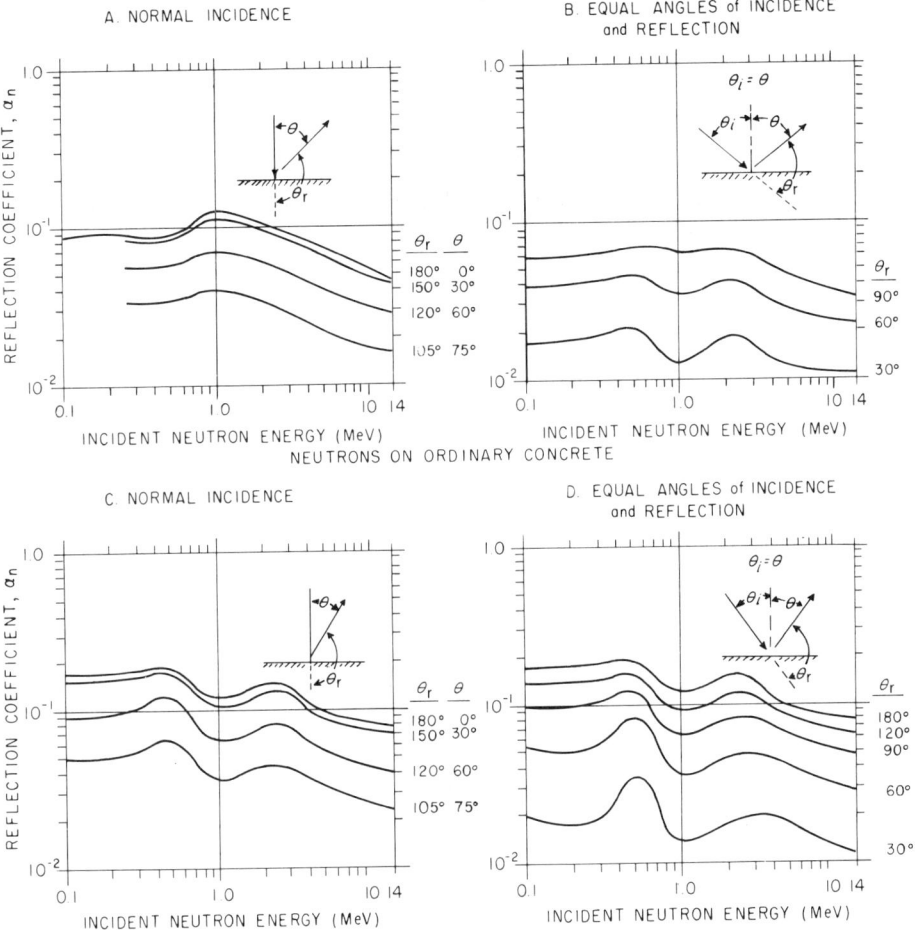

Reflection coefficients, α_n, for monoenergetic fast neutrons on ordinary concrete, iron and lead, as a function of incident monoenergetic neutron energy, for several angles of reflection, assuming normal incidence as well as equal angles of incidence and reflection. Values of α_n are given for neutron incidence on ordinary concrete and iron. References are given in Appendix E-15.

Values for α_n for fast neutrons incident on lead are probably an order of magnitude higher than those given above, indicating that thick lead barriers are not desirable for capturing neutrons. This conclusion is derived from the albedo formula of Chandrasekhar (1960), with the assumption that, for fast neutrons, the scattering by individual atoms is almost isotropic and that the absorption cross section is comparatively small. This assumption should give an upper limit to the actual values of α_n for lead.

The values of α_n for $\theta_r = 180°$ in Curve A are the same as for $\theta_r = 180°$ in Curve B.

APPENDIX G

Radioactivity

G.1 Radioactivity-Producing Nuclear Reactions

Reference is made to Barbier (1969) for a comprehensive treatment of this subject.
Typical photoneutron reactions are listed below.

Target	(γ, n) Threshold energy MeV	Product	Product half-life
^{12}C	18.7	^{11}C	20.5 m
^{14}N	10.5	^{13}N	10.0 m
^{16}Q	15.7	^{15}O	124. s
^{27}Al	13.1	^{26}Al	6.5 s
^{54}Fe	13.6	^{53}Fe	8.5 m
^{65}Cu	9.9	^{64}Cu	12.9 h
^{70}Zn	9.2	^{69}Zn	52.0 m
^{82}Se	9.8	^{81}Se	17.0 m
^{107}Ag	9.5	^{106}Ag	24. m/8.3 d
^{115}In	9.0	^{114}In	72. s/50. d
^{121}Sb	9.25	^{120}Sb	17.0 m
^{127}I	9.1	^{126}I	12.8 d
^{141}Pr	9.4	^{140}Pr	3.6 m
^{181}Ta	7.6	^{180}Ta	8.1 h
^{182}W	8.0	^{181}W	130. d
^{197}Au	8.1	^{196}Au	6.2 d/9.7 h
^{204}Pb	8.2	^{203}Pb	6.1 s/52. h
^{2}H	2.20	^{1}H	stable
^{9}Be	1.67	^{8}Be $\rightarrow$ 2 ^{4}He	stable

G.2 Induced Radioactivity in Cyclotrons

1. Decay curves for gamma radiation following shutdown of the University of California 60-inch cyclotron after operation with 24-MeV deuterons are shown in Appendix G-3. The primary half-lives observed are 5 minutes, 38 minutes, 2.6 hours, and 12.8 hours. These are probably radiations from ^{66}Cu, ^{65}Ni and/or ^{56}Mn and ^{64}Cu respectively (Reference 2 in Brobeck and Associates, 1968). It is apparent that it would be wise to delay maintenance operations on this cyclotron for at least 8 hours and preferably over a weekend.
2. The major long-lived radioactivity in moderate-energy cyclotrons (22 MeV protons) is ^{65}Zn in the dee structures. Under specialized circumstances, radioactivity is found concentrated in the tungsten beam-extraction septum. Aluminum collimating slits are used when neutron production from them does not interfere; otherwise tantalum slits are used (Hendry, 1972).
3. Listed below are typical components that become activated at the Texas A&M Variable Energy Cyclotron (TAMVEC). Activation of beam-line components from fast neutrons is also a significant problem (Kreger, 1971).

Component	Radionuclide	Activity mCi
Cyclotron probes	^{65}Zn, ^{64}Cu	~30
Deflector	^{182}Ta, ^{52}Mn (^{56}Co, ^{64}Cu)	~900
Faraday cups	^{13}N, ^{11}C	~120
Collimating slits	^{182}Ta, (^{24}Na)	~110
View plates	^{24}Na (^{24}Al)	~10

G.3 Gamma-Radiation Dose-Equivalent Index Rate Following Cyclotron Shutdown

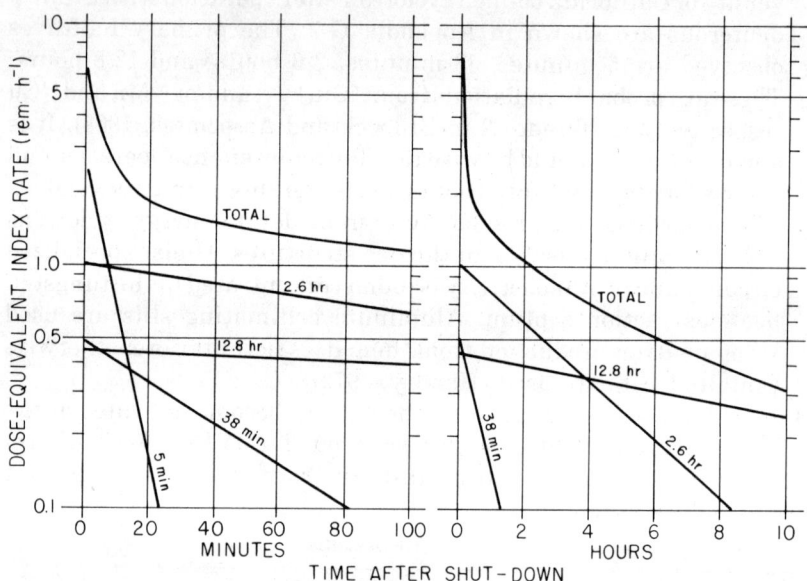

Gamma-radiation dose-equivalent index rate as a function of time after cyclotron shutdown, as measured beside the vacuum tank of the University of California 60-inch cyclotron (from Brobeck and Associates, 1968).

APPENDIX H

Shielding Materials

H.1 Densities of X-Ray Shielding Materials

Material	Average density	
	g cm^{-3}	lb ft^{-3}
Concrete		
ordinary	2.35	147
barytes	3.6	210
ferrophosphorus aggregate	4.8	300
ilmenite aggregate	3.85	220
Earth, dry, packed	1.5	95
Sand, dry, packed	1.6–1.9	100–120
Brick		
soft	1.65	103
hard	2.05	128
Tile, ceramic	1.9	118
Sand plaster	1.54	96
Granite	2.65	165
Limestone	2.46	153
Marble	2.7	170
Barium sulfate (natural barite)	4.5	280
Air (NTP)	1.293×10^{-3}	8.072×10^{-2}
Water	1.0	62.4
Wood (construction use)	0.5–0.9	31–56
Lead glass		
ordinary	3.27	205
high-density	6.22	387
Copper	8.9	556
Steel	7.8	489
Lead	11.36	709

Note: Concrete blocks and cinder blocks vary too much to be listed.

Information obtained from: (1) NCRP Report No. 49 (NCRP, 1976); (2) Etherington (1958); (3) Walker and Grotenhuis (1961); and (4) Hodgman (1952).

H.2 Neutron Shielding Materials: Elemental Composition of Concretes[a]

Element	TSF-5.5	TSF-3.0	TSF-8.0	NBS-5.0
H	8.50	4.64	12.36	7.76
C	20.20	20.73	19.67	–
O	35.50	34.395	36.605	43.29
Mg	1.86	1.91	1.81	1.17
Al	0.60	0.62	0.58	2.35
Si	1.70	1.74	1.66	15.68
Ca	11.30	11.60	11.00	3.55
Fe	0.19	0.20	0.18	0.303
Percent water, by weight	5.5	3.0	8.0	5.0
Density (g cm^{-3})	2.31	2.31	2.31	2.31

[a] Concentrations are in units of 10^{21} atoms cm^{-3}.

TSF refers to compositions established for the concrete used at the Tower Shielding Facility at the Oak Ridge National Laboratory (data obtained from Chilton, 1971).

NBS refers to composition closely similar to the concrete usually used in studies at the National Bureau of Standards, sometimes known as "Grodstein-McGinnies" concrete (data obtained from Chilton, 1971). The primary difference in all these concretes is a slight reduction in density from the standard value of 2.35.

Note: See Appendix H-3 for guidance on the effects of water content and elemental composition on neutron-attenuation properties of concretes.

H.3 Neutron Shielding Materials: Effects of Water Content and Elemental Composition on Neutron-Transmission Properties of Concrete

Note: Neutron-transmission curves in this report (Appendices F-6, F-8, F-9) are based on TSF-5.5 carbonaceous type of concrete with 5.5 percent water content.

1. *Effect of Water Content:* If the water content of a given concrete is less than 5.5 percent by weight, the following multiplying factor should be applied to the dose-equivalent index transmission per unit fluence of the neutron-transmission curves in this report:

Water Content percent	Multiplying Factor
2.5	4.1
3.5	2.6
4.5	1.6
5.5	1.0
>5.5	1.0

   This factor is most closely valid for neutrons in the low-MeV range (1 to 15 MeV), and for shield thicknesses about 400 g cm^{-2}. For other thicknesses and energies, the factor is less. For neutrons above about 25 MeV, variation in water content (within reasonable limits) has little effect. In case of doubt about the water content, the lowest reasonable value should be assumed.

2. *Effect of Aggregate:* Siliceous types of concrete (e.g. NBS-5.0, see Appendix H-2) are not as effective as carbonaceous types for attenuating neutrons, for the same thickness (in terms of g cm^{-2}) and for the same water content. In fact, siliceous concretes exhibit characteristics like carbonaceous concretes having a water content 1 to 2 percent lower, i.e. a 5 percent siliceous concrete behaves like a carbonaceous concrete with 3 or 4 percent water content, for the same mass thickness. The table above can be used to correct the dose-equivalent transmission of the neutron-transmission curves in this report.

Reference: Chilton (1971).

H.4 Neutron Shielding Materials: Miscellaneous Materials

Reference is made to Appendix E of NCRP Report No. 38 (NCRP, 1971a), for a comparison of monoenergetic-neutron transmission through ordinary concrete, water, polyethylene, and soil. A calculation method is described for estimating shielding thicknesses of other materials, based on neutron-removal cross sections for fission neutrons or similar spectra that peak at a few MeV. These calculations are suitable for thin materials, in which the production of neutron capture gamma rays is relatively unimportant.

APPENDIX I

Ozone and Other Noxious Gases

I.1 Production of Ozone by External Electron Beams

Assumptions:
No disassociation of ozone (O_3) during period of irradiation;
No ventilation of the room in which the external electron beam is contained;
O_3 is distributed uniformly throughout the volume of the room during irradiation.

The ozone concentration in air, C_{O_3}, after an irradiation period, t, is given by

$$C_{O_3} = \left[\frac{C_{O_2} G \epsilon}{N}\right] \left[\frac{S_{col} I \, x \, t}{v}\right] \times 10^6$$

$$= 3.25 \left[\frac{S_{col} I \, x \, t}{v}\right] \text{ ppm}$$

where
- C_{O_2} is the fractional concentration of oxygen in air (0.232);
- G is the G-value for O_3 production by electron irradiation of oxygen (~6 molecules per 100 electron-volts[a]);
- ϵ is the number of electronic charges per milliampere-second of electron-beam current (6.28×10^{15} mA^{-1} s^{-1});
- N is Avogadro's Number (6.02×10^{23} molecules per 22.4 liters of gas at NTP);
- S_{col} is the collision stopping power of electrons in air at NTP (keV cm^{-1})[b];
- I is the external electron-beam current (mA);
- x is the distance in air traversed by the external electron beam (cm);
- t is the irradiation period (s);

and
- v is the volume of the room containing the external electron beam (l).

As an example of ozone production, a 1-mA external beam of 10-MeV electrons, traversing a 300-cm distance in a room 4 m × 4 m × 3 m (about 5×10^4 l), can produce an ozone concentration of about 15 ppm after a 5-minute irradiation period. Reference is made to Appendix I-2 for threshold limit values for ozone.

[a] Frequently reported G-values are between 3 and 9. For the purpose of this report, a value of 6 is used (Brynjolfsson and Martin, 1967).

[b] Typical values for S_{col}:

0.5 MeV	2.3 keV cm^{-1}
1.0 MeV	2.2 keV cm^{-1}
10 MeV	2.5 keV cm^{-1}
50 MeV	3.0 keV cm^{-1}

I.2 Threshold Limit Values for Ozone and Certain Oxides of Nitrogen in Workroom Air[a]

Substance	Threshold Limit Values (TLV)[a] ppm
Ozone	0.1
Nitric oxide	25
Nitrogen dioxide	5

According to ACGIH (1971b), ozone is a highly injurious and lethal gas at relatively low concentrations (a few ppm) and at short exposure periods (a few hours). Ozone is apparently radiomimetic, i.e. it mimics, in its effects on health, those of ionizing radiation. The TLV of 0.1 ppm for ozone represents a limit which, although it results in no ostensible or manifest injury, may result in premature aging in a manner similar to that from continued exposure to ionizing radiation, if exposure is sufficiently prolonged.

If electron-beam exposures to the air are prolonged (as for electron-beam processing), it may be necessary to provide for very high ventilation for the ozone-contaminated volume of air. Facilities for electron-beam generators *should* be designed so that the path of the electron beam in air is short. A high-capacity air exhaust system *should* preferably be close to the beam output.

After machine shutoff, the concentration of ozone in the affected volume of air is given by:

$$C'_{O_3} = C_{O_3} e^{-\left(\frac{f t'}{v}\right)}$$

where

C'_{O_3} is ozone concentration (ppm) at time t'
f is pump exhaust rate (liters s^{-1})
t' is time after shutoff (s).

[a] ACGIH (1971a).

References

ACGIH (1971a). American Conference of Governmental Industrial Hygienists, *Threshold Limit Values of Airborne Contaminants and Physical Agents with Intended Changes Adopted by ACGIH for 1971* (American Conference of Governmental Industrial Hygienists, Cincinnati).

ACGIH (1971b). American Conference of Governmental Industrial Hygienists, *Documentation of the Threshold Limit Values for Substances in Workroom Air*, 3rd ed. (American Conference of Governmental Industrial Hygienists, Cincinnati).

ADAMS, G. D. AND GIRARD, J. P. (1946). "Application of the betatron to practical radiography," Trans. Am. Inst. Electr. Eng. 65, 241.

ALSMILLER, R. G., JR., ARMSTRONG, T. W. AND COLEMAN, W. A. (1970). *The Absorbed Dose and Dose Equivalent from Neutrons in the Energy Range 60 to 3000 MeV and Protons in the Energy Range 400 to 3000 MeV*, Report ORNL-TM-2924 (rev.) (Oak Ridge National Laboratory, Oak Ridge, Tennessee).

AMOS, T., GALONSKY, A. AND JOLLY, R. (1970). "Neutron yields from proton bombardment of thick targets," Bull. Am. Phys. Soc., Series II, 15, No. 10, 1692.

ANSI (1971). American National Standards Institute, *Marking Physical Hazards and the Identification of Certain Equipment (Safety Color Code)*, ANSI Z53.1-1971 (American National Standards Institute, New York).

AUXIER, J. A., SNYDER, W. S. AND JONES, T. D. (1968). "Neutron interactions and penetration in tissue," page 275 in *Radiation Dosimetry*, Vol. I, 2nd ed., Attix, F. H. And Roesch, W. C., Eds. (Academic Press, New York).

AXTON, E. J. AND BARDELL, A. G. (1972). "Neutron production from electron accelerators used for medical purposes," Phys. Med. Biol. 17, 293.

BARBIER, M. (1969). *Induced Radioactivity* (John Wiley & Sons, New York).

BARKAS, W. H. AND BERGER, M. J. (1964). *Tables of Energy Losses and Ranges of Heavy Charged Particles*, NASA Report No. SP-3013 (National Aeronautics and Space Agency, Washington).

BERGER, M. J. AND SELTZER, S. M. (1970). "Bremsstrahlung and photoneutrons from thick tungsten and tantalum targets," Phys. Rev. C. 2, 621.

BLY, J. H. (1964). "High energy radiography 1–30 MeV," Mater. Eval. 22, 519.

BLY, J. H. AND BURRILL, E. A. (1959). "High-energy radiography in the 6- to 30-MeV range, " ASTM Spec. Tech. Pub. No. 278, 20.

BOLT, R. O. AND CAROLL, J. G. (1963). *Radiation Effects on Organic Materials* (Academic Press, New York).

BORCHERS, R. R. AND WOOD, R. M. (1965). "Neutrons from the deuteron bombardment of thick targets," Nucl. Instrum. Methods 35, 138.

BROBECK, W. M. AND ASSOCIATES (1968). *Particle Accelerator Safety Manual*,

Public Health Service Report No. MORP 68-12 (U.S. Department of Health, Education and Welfare, Washington).
BROMLEY, D. A., FERGUSON, A. J., GOVE, H. E., KUEHNER, J. A., LITHERLAND, A. E., ALMQVIST, E. AND BATCHELOR, R. (1959). "Studies of (p,n) reactions in the proton energy range from 2-10 MeV," Can. J. Phys. 37, 1514.
BRUNINX, E. AND CROMBEEN, J. (1969). "Thick target neutron yields and neutron spectra produced by 20 MeV helium-3 ions, 14 MeV protons and 7.5 MeV deuterons on a beryllium target," Int. J. Appl. Radiat. Isot. 20, 255.
BRYNJOLFSSON, A. AND MARTIN, T. G., III (1967). "Radiation protection problems associated with electron accelerators," Radiation Protection Symposium (U.S. Army, Edgewood Arsenal, Maryland).
BUECHNER, W. W., VAN DE GRAAFF, R. J., FESHBACH, H., BURRILL, E. A., SPERDUTO, A. AND McINTOSH, L. R. (1948a). "An investigation of radiography in the range from 0.5 to 2.5 million volts," ASTM Bull. No. 155, 54.
BUECHNER, W. W., VAN DE GRAAFF, R. J., BURRILL, E. A. AND SPERDUTO, A. (1948b). "Thick-target x-ray production in the range 1250 to 2350 kV," Phys. Rev. 74, 1348.
BURRILL, E. A. (1963). *Neutron Production and Protection* (High Voltage Engineering Corp., Burlington, Massachusetts).
BURRILL, E. A. (1965). *Radiation Shielding Considerations for Tandem Accelerators*, Publication ANS-SD-3 (American Nuclear Society, Hinsdale, Illinois).
BURRILL, E. A. (1966). *Analysis of X-Ray Profiles from Van de Graaff Ion Accelerators*, Internal Memorandum (High Voltage Engineering Corp., Burlington, Massachusetts).
BURRILL, E. A. (1968). "Shielding of direct electron accelerators," page 135 in *Engineering Compendium on Radiation Shielding*, Vol. III, Jaeger, R. G., Ed-in-ch. (Springer-Verlag, New York).
BURRILL, E. A. (1972). "Radioisotopes—or particle accelerators?" page 361 in *Radioisotope Engineering*, Eichholz, G., Ed. (Marcel Dekker, Inc., New York).
CHANDRASEKHAR, S. (1960). *Radiative Transfer* (Dover Publ. Co., New York).
CHILTON, A. B. (1964). *Backscatter for Gamma Rays from a Point Source Near a Concrete Plane Surface*, Univ. Illinois Eng. Sta. Bull. No. 471 (University of Illinois, Urbana).
CHILTON, A. B. (1965). "Backscattering of gamma rays from point sources by an infinite-plane concrete surface," Nucl. Sci. Eng. 21, 194.
CHILTON, A. B. (1971). *Effect of Material Composition on Neutron Penetration of Concrete Slabs*, NBS Report No. 10425 (National Bureau of Standards, Washington).
CHILTON, A. B. (1974). *Working Paper on Neutron Skyshine for Accelerator Sources*, Informal report (University of Illinois, Urbana).
CHILTON, A. B. AND HUDDLESTON, C. M. (1963). "A semi-empirical formula for differential dose albedo for gamma rays on concrete," Nucl. Sci. Eng. 17, 419.
CHILTON, A. B. AND MORRIS, E. E. (1972). "Shielding effectiveness of ribbed

slabs against gamma radiation: Part II. Engineering methods," Nucl. Eng. Des. **23**, 367.

CHILTON, A. B., DAVISSON, C. M. AND BEACH, L. A. (1965). "Parameters for C-H albedo formula for gamma rays reflected from water, concrete, iron, and lead," Trans. Am. Nucl. Soc. **8**, No. 2, 656.

CLARKE, E. T. (1968). "Photon fields near earth-air interface," page 255 in *Engineering Compendium on Radiation Shielding*, Vol. I, Jaeger, R. G., Ed-in-ch. (Springer-Verlag, New York).

CLELAND, M. (1961). "Megaroentgen dose rates," page 1 in *Radiation Review*, Vol. 1, No. 3 (Radiation Dynamics Inc., Westbury, New York).

DASA (1969). Defense Atomic Support Agency, *Transient Radiation Effects on Electronics Handbook*, 2nd ed., DASA-1420 (Defense Atomic Support Agency, Washington).

DNA (1972). Defense Nuclear Agency, *TREE Preferred Procedures*, 2nd ed., DNA-2028H (Defense Nuclear Agency, Washington).

ETHERINGTON, H., Ed. (1958). *Nuclear Engineering Handbook* (McGraw-Hill Book Co., Inc., New York).

FIRK, F. W. K. (1967). "Energy spectra of photoneutrons at excitation energies up to 60 MeV," page 352 in *Proceedings of the International Physics Conference, Gatlinburg, Tenn.* (Academic Press, New York).

FLEISCHER, A. A. (1968). "The production of fast neutrons by small cyclotrons," TCC Report 2003 (The Cyclotron Corporation, Berkeley, California); Symposium on Fast Neutron Therapy, Radiation Research Society, Houston, Texas, April 25, 1968.

FOWLER, J. L. AND BROLLEY, J. E. (1956). "Monoenergetic neutron techniques in the 10-30 MeV range," Rev. Mod. Phys. **28**, 103.

FRANK, H. (1959). "Zur Vielfachstreuung und Rückdiffusion Schneller Electronen nach Durchgang durch dicks Schlichten," Z. Naturforsch. **14a**, 247.

GLAZUNOV, Y. Y., SAVIN, M. V., SAFINA, I. N., FOMUSKIN, E. F. AND KHOKHLOV, Y. A. (1964). "Photoneutron spectra of platinum, bismuth, lead, and uranium," Sov. Phys., JETP **19**, 1284.

GOLDIE, C. H., WRIGHT, K. A., ANSON, J. H., CLOUD, R. W. AND TRUMP, J. G. (1954). "Radiographic properties of x rays in the 2-6 MeV range," ASTM Bull. No. 201, 49.

GOVE, H. E. (1956). *AECL Chalk River Estimates of Radiation Protection* (Atomic Energy of Canada, Ltd., Chalk River, Ontario).

GOZANI, T. AND KULL, L. A. (1969). *Neutron Production in ^{238}U and ^{235}U by Low Energy Electron Bremsstrahlung*, Report No. GA-9717 (Gulf General Atomic, Inc., San Diego, California).

GREEN, D. W., MORRIS, E. E., PREISS, K. AND CHILTON, A. B. (1972). "Shielding effectiveness of ribbed slabs against gamma radiation: Part I. Experimental and theoretical studies," Nucl. Eng. Des. **23**, 359.

HANSON, A. O. AND TASCHEK, R. F. (1950). *Monoenergetic Neutrons from Charged Particle Reactions*, Preliminary Report No. 4, Nuclear Science Series (National Research Council, Washington).

HARDER, D. (1965). *Transmission of Fast Electrons Through Thick Materials* (Habilitationsschrift, Julius-Maximilians University, Wurtzburg, Germany).

HENDRY, G. O. (1972). Personal communication (Cyclotron Corp., Berkeley, California).
HODGMAN, C. D., Ed-in-ch. (1952). *Handbook of Chemistry and Physics* (Chemical Rubber Publishing Co., Cleveland).
HOLEMAN, G. R., SHAW, D. M. AND PRICE, K. W. (1969). "Stray neutron spectra and comparison of measurements with discrete ordinate calculations," page 552 in *Proceedings of the Second International Conference on Accelerator Dosimetry and Experience*, Report No. CONF-691101 (National Technical Information Service, Springfield, Virginia).
IAEA (1961). International Atomic Energy Agency, *Radioactive Waste Disposal into the Sea*, IAEA Report No. 5, Safety Series (International Atomic Energy Agency, Vienna).
IAEA (1965). International Atomic Energy Agency, *Radioactive Waste Disposal into the Ground*, IAEA Report No. 15, Safety Series (International Atomic Energy Agency, Vienna).
IAEA (1967). International Atomic Energy Agency, *Basic Factors for the Treatment and Disposal of Radioactive Wastes*, IAEA Report No. 24, Safety Series (International Atomic Energy Agency, Vienna).
IAEA (1971). International Atomic Energy Agency, *Disposal of Radioactive Wastes into Rivers, Lakes, and Estuaries*, IAEA Report No. 36, Safety Series (International Atomic Energy Agency, Vienna).
ICRP (1959). International Commission on Radiological Protection, *Report of Committee II on Permissible Dose for Internal Radiation*, ICRP Publication 2 (Pergamon Press, London).
ICRP (1964). International Commission on Radiological Protection, *Protection Against Electromagnetic Radiations Above 3 MeV, and Electrons, Neutrons and Protons*, ICRP Publication 4 (International Commission on Radiological Protection, Macmillan Co., New York).
ICRP (1971). International Commission on Radiological Protection, *Report of Committee 3 on Data for Protection Against Ionizing Radiation from External Sources: Supplement to ICRP Publication 15*, ICRP Publication 21 (Pergamon Press, London).
ICRU (1969). International Commission on Radiation Units and Measurements, *Neutron Fluence, Neutron Spectra and Kerma*, ICRU Report 13 (International Commission on Radiation Units and Measurements, Washington).
ICRU (1971a). International Commission on Radiation Units and Measurements, *Radiation Quantities and Units*, ICRU Report 19 (International Commission on Radiation Units and Measurements, Washington).
ICRU (1971b). International Commission on Radiation Units and Measurements, *Radiation Protection Instrumentation and Its Application*, ICRU Report 20 (International Commission on Radiation Units and Measurements, Washington).
ICRU (1973). International Commission on Radiation Units and Measurements, *Dose Equivalent*, Supplement to ICRU Report 19 (International Commission on Radiation Units and Measurements, Washington).
IRVING, D. C., ALSMILLER, R. G., JR. AND MORAN, H. S. (1967). *Tissue*

Current-to-Dose Conversion Factors for Neutrons with Energies from 0.5 to 60 MeV, Report ORNL-4032 (Oak Ridge National Laboratory, Oak Ridge, Tennessee).

KARZMARK, C. J. AND CAPONE, T. (1968). "Measurements of 6 MV x rays: 1. Primary radiation absorption in lead, steel, and concrete," Br. J. Radiol. 41, 33.

KATZ, L. AND PENFOLD, A. S. (1952). "Range-energy relations for electrons and the determination of beta-ray endpoint energies by absorption," Rev. Mod. Phys. 24, 28.

KAUSHAL, N. N., WINHOLD, E. J., YERGIN, P. F., MEDICUS, H. A. AND AUGUSTON, R. H. (1968). "Fast photoneutron spectra due to 50 to 85 MeV photons," Phys. Rev. 175, 1330.

KIRCHER, J. AND BOWMAN, R. E., Eds. (1964). *Effects of Radiation on Materials and Components* (Reinhold Publ. Co., New York).

KIRN, F. S. AND KENNEDY, R. J. (1954). "How much concrete for shielding: betatron x rays," Nucleonics 12, No. 6, 44.

KOHL, J., ZENTNER, R. D. AND LUKENS, H. R. (1961). Chapter 14, "Process uses of radiation," page 411 in *Radioisotope Applications Engineering*, (D. Van Nostrand Co., Inc., Princeton).

KREGER, W. E. (1971). Informal report (Physics International Co., San Leandro, California).

LADU, M., PELLICCIONI, M., PICCHI P. AND VERRI, G. (1968). "A contribution to the skyshine study," Nucl. Instrum. Methods 62, 51.

LINDENBAUM, S. T. (1957). "Shielding of high-energy accelerators," *Conference on Shielding of High-Energy Accelerators*, Report No. USAEC TID-7545 (U.S. Atomic Energy Commission, Washington).

LOEVINGER, R., KARZMARK, C. J. AND WEISSBLUTH, M. (1961). "Radiation therapy with high-energy electrons," Radiology 77, 906.

LOKAN, K. H., SHERMAN, N. K., GELLIE, R. W., HENRY, W. H., LEVESQUE, R., NOWAK, A. AND TEATHER, G. C. (1972). "Bremsstrahlung attenuation measurements in ilmenite loaded concretes," Health Phys. 23, 193.

MACGREGOR, M. H. (1959). *Neutrons from Photodisintegration of Beryllium*, Report AM-109 (Applied Radiation Corp., Walnut Creek, California).

MAERKER, R. E. AND MUCKENTHALER, F. J. (1967). "Neutron fluxes in concrete ducts arising from incident epicadmium neutrons: Calculations and experiments," Nucl. Sci. Eng. 30, 340.

MAERKER, R. E., CLAIBORNE, H. C. AND CLIFFORD, C. E. (1968). "Neutron attenuation in rectangular ducts," page 517 in *Engineering Compendium on Radiation Shielding*, Vol. I, Jaeger, R. G., Ed-in-ch. (Springer-Verlag, New York).

MARION, J. B. AND FOWLER, J. L. (1960). *Fast Neutron Physics, Part I* (Interscience Publishers, New York).

MARUYAMA, T., KUMAMOTO, Y., KATO, Y. AND YAMAMOTO, M. (1971). "Attenuation of 4–32 MeV x-rays in ordinary concrete, heavy concrete, iron, and lead," Health Phys. 20, 277.

MCCALL, R. C. AND NELSON, W. R. (1974). Personal communication (Stanford Linear Accelerator Center, Stanford, California).

REFERENCES / 141

McCaslin, J. B. and Smith, A. R. (1969). "Radiation measurements and shielding study of the Berkeley 27-inch ³He cyclotron," page 486 in *Proceedings of the Second International Conference on Accelerator Dosimetry and Experience,* Report No. CONF-691101 (USAEC Div. Tech. Info., Stanford, California).

McMaster, R. C., Ed. (1963). "Radiation protection," page 26.1 in *Nondestructive Testing Handbook,* Vol. 1 (Ronald Press, New York).

Miller, C. W. (1954). "Industrial radiography and the linear accelerator," J. Br. Inst. Radio Eng. 14, No. 8, 361.

Miller, W. and Kennedy, R. J. (1956). "Attenuation of 86 and 176 MeV synchrotron x-rays in concrete and lead," Radiat. Res. 4, 360.

NCRP (1951a). National Council on Radiation Protection and Measurements, *Control and Removal of Radioactive Contamination in Laboratories,* NCRP Report No. 8, published as National Bureau of Standards Handbook 48 (U.S. Government Printing Office, Washington).

NCRP (1951b). National Council on Radiation Protection and Measurements, *Recommendations for Waste Disposal of Phosphorus-32 and Iodine-131 for Medical Users,* NCRP Report No. 9, published as National Bureau of Standards Handbook 49 (U.S. Government Printing Office, Washington).

NCRP (1954a). National Council on Radiation Protection and Measurements, *Protection Against Betatron-Synchrotron Radiations up to 100 Million Electron Volts,* NCRP Report No. 14, published as National Bureau of Standards Handbook 55 (U.S. Government Printing Office, Washington).

NCRP (1954b). National Council on Radiation Protection and Measurements, *Radioactive Waste Disposal in the Ocean,* NCRP Report No. 16, published as National Bureau of Standards Handbook 58 (U.S. Government Printing Office, Washington).

NCRP (1957). National Council on Radiation Protection and Measurements, *Protection Against Neutron Radiation Up to 30 Million Electron Volts,* NCRP Report No. 20, published as National Bureau of Standards Handbook 63 (U.S. Government Printing Office, Washington).

NCRP (1959). National Council on Radiation Protection and Measurements, *Maximum Permissible Body Burdens and Maximum Permissible Concentrations of Radionuclides in Air and Water for Occupational Exposure,* NCRP Report No. 22, published as National Bureau of Standards Handbook 69 (U.S. Government Printing Office, Washington).

NCRP (1964a). National Council on Radiation Protection and Measurements, *Safe Handling of Radioactive Materials,* NCRP Report No. 30, published as National Bureau of Standards Handbook 92 (U.S. Government Printing Office, Washington).

NCRP (1964b). National Council on Radiation Protection and Measurements, *Shielding for High-Energy Electron Accelerator Installations,* NCRP Report No. 31, published as National Bureau of Standards Handbook 97 (U.S. Government Printing Office, Washington).

NCRP (1970a).[11] National Council on Radiation Protection and Measurements, *Medical X-Ray and Gamma-Ray Protection for Energies up to 10 MeV – Structural Shielding Design and Evaluation,* NCRP Report No. 34

(National Council on Radiation Protection and Measurements, Washington).
NCRP (1970). National Council on Radiation Protection and Measurements, *Dental X-Ray Protection,* NCRP Report No. 35 (National Council on Radiation Protection and Measurements, Washington).
NCRP (1971a). National Council on Radiation Protection and Measurements, *Protection Against Neutron Radiation,* NCRP Report No. 38 (National Council on Radiation Protection and Measurements, Washington).
NCRP (1971b). National Council on Radiation Protection and Measurements, *Basic Radiation Protection Criteria,* NCRP Report No. 39 (National Council on Radiation Protection and Measurements, Washington).
NCRP (1976). National Council on Radiation Protection and Measurements, *Structural Shielding Design and Evaluation for Medical Use of X Rays and Gamma Rays of Energies Up to 10 MeV,* NCRP Report No. 49 (National Council on Radiation Protection and Measurements, Washington).
NCRP (1977). National Council on Radiation Protection and Measurements, *Radiation Protection for Small Neutron Generators,* in preparation (National Council on Radiation Protection and Measurements, Washington).
NELLIS, D. O., HUDSPETH, E. L., MORGAN, I. L., BUCHANAN, P. S. AND BOGGS, R. F. (1967). *Tritium Contamination in Particle Accelerator Operation,* Public Health Service Publication No. 999-RH-29 (U.S. Department of Health, Education and Welfare, Washington).
O'CONNOR, D. T., CRISCUOLO, E. L. AND PACE, A. L. (1949). "10-MeV x-ray technique," page 1 in *Papers on Radiography,* ASTM Spec. Tech. Pub. No. 96, 10.
ROUSSIN, R. W. AND SCHMIDT, F. A. R. (1971). "Adjoint S_N calculations of coupled neutron and gamma-ray transport through concrete slabs," Nucl. Eng. Des. 15, No. 3, 319.
ROUSSIN, R. W., ALSMILLER, R. G., JR. AND BARISH, J. (1973). "Calculations of the transport of neutrons and secondary gamma rays through concrete for incident neutrons in the energy range 15 to 75 MeV," Nucl. Eng. Des. 24, 250.
SAXON, G. (1964). "Radiation protection aspects of the design and operation of irradiation facilities," page 200 in *Radiation Sources,* Charlesby, A., Ed. (Pergamon Press, London).
SCAG, D. A. (1954). "Discussion of radiographic characteristics of high energy x-rays," Nondestr. Test. 12, No. 3, 47.
SCHWEIMER, G. W. (1967). "Fast neutron production with 54-MeV deuterons," Nucl. Phys. A100, 537.
SHPETNYI, A. I. (1957). "Energy and angular distribution of neutrons emitted in the $^9Be(d,n)^{10}B$ reaction," J. Exptl. Theor. Phys. USSR, 32, 423; Trans. Sov. Phys., JETP 5, 357.
SPINKS, J. W. T. AND WOODS, R. J. (1964). *An Introduction to Radiation Chemistry* (John Wiley and Sons, New York).
STEPHENS, L. D. AND MILLER, A. J. (1969). "Radiation studies at a medium energy accelerator," page 459 in *Proceedings of the Second International Conference on Accelerator Dosimetry and Experience,* Report No. CONF-691101 (National Technical Information Service, Springfield, Virginia).

TRUMP, J. G. AND VAN DE GRAAFF, R. J. (1949). "The secondary emission of electrons by high energy electrons," Phys. Rev. 75, 44.

TRUMP, J. G., VAN DE GRAAFF, R. J. AND CLOUD, R. W. (1940). "Cathode rays for radiation therapy," Am. J. Roentgenol. Radium Ther. 49, 728.

TRUMP, J. G., WRIGHT, K. A. AND CLARKE, A. M. (1950). "Distribution of ionization in materials irradiated by two- and three-million-volt cathode rays," J. Appl. Phys. 21, 345.

VERBINSKI, V. V. AND COURTNEY, J. C. (1965). "Photoneutron spectra and cross sections for ^{12}C and ^{16}O," Nucl. Phys. 73, 398.

WALKER, R. L. AND GROTENHUIS, M. (1961). *A Summary of Shielding Constants for Concrete*, ANL Report No. 6443 (Argonne National Laboratory, Argonne, Illinois).

WESTENDORP, W. F. AND CHARLTON, E. D. (1945). "A 100-MeV induction electron accelerator," J. Appl. Phys. 16, 581.

WIDEROË, R. (1953). "The Brown-Boveri 31-MeV dual beam betatron," Nondestr. Test. 11, No. 4, 23.

WRIGHT, H. A., ANDERSON, U. E., TURNER, J. E., NEUFELD, J. AND SNYDER, W. S. (1969). "Calculation of radiation dose due to protons and neutrons with energies from 0.4 to 2.4 GeV," Health Phys. 16, 13.

WRIGHT, K. A. AND TRUMP, J. G. (1962). "Back-scattering of megavolt electrons from thick targets," J. Appl. Phys. 33, 687.

WYCKOFF, J. M. AND CHILTON, A. B. (1973). "Dose due to practical neutron energy incident on concrete shielding walls," page 694 in *Proceedings of the Third International Congress of the International Radiation Protection Association*, Snyder, W. S., Ed., Report No. CONF-730907 (National Technical Information Service, Springfield, Virginia).

WYCKOFF, J. M., PRUITT, J. S. AND SVENSSON, G. (1971). "Dose vs. angle and depth produced by 20 to 100 MeV electrons incident on thick targets," page 773 in the *Proceedings of the International Congress on Protection Against Accelerator and Space Radiations*, Vol. 2, Report No. Cern-71-16 (Cern, Geneva, Switzerland).

ZERBY, C. D. AND KINNEY, W. E. (1965). "Calculated tissue current-to-dose conversion factors for nucleons below 400 MeV," Nucl. Instrum. Methods 36, 125.

The NCRP

The National Council on Radiation Protection and Measurements is a nonprofit corporation chartered by Congress in 1964 to:
1. Collect, analyze, develop, and disseminate in the public interest information and recommendations about (a) protection against radiation and (b) radiation measurements, quantities, and units, particularly those concerned with radiation protection;
2. Provide a means by which organizations concerned with the scientific and related aspects of radiation protection and of radiation quantities, units, and measurements may cooperate for effective utilization of their combined resources, and to stimulate the work of such organizations;
3. Develop basic concepts about radiation quantities, units, and measurements, about the application of these concepts, and about radiation protection;
4. Cooperate with the International Commission on Radiological Protection, the International Commission on Radiation Units and Measurements, and other national and international organizations, governmental and private, concerned with radiation quantities, units, and measurements and with radiation protection.

The Council is the successor to the unincorporated association of scientists known as the National Committee on Radiation Protection and Measurements and was formed to carry on the work begun by the Committee.

The Council is made up of the members and the participants who serve on the fifty-four Scientific Committees of the Council. The Scientific Committees, composed of experts having detailed knowledge and competence in the particular area of the Committee's interest, draft proposed recommendations. These are then submitted to the full membership of the Council for careful review and approval before being published.

The following comprise the current officers and membership of the Council:

Officers

President	Lauriston S. Taylor
President-Elect	Warren K. Sinclair
Vice President	E. Dale Trout
Secretary and Treasurer	W. Roger Ney
Assistant Secretary	Eugene R. Fidell
Assistant Treasurer	Harold O. Wyckoff

Members

Seymour Abrahamson
Roy E. Albert
John A. Auxier
William J. Bair
Victor P. Bond
Harold S. Boyne
Robert L. Brent
A. Bertrand Brill
Reynold F. Brown
William W. Burr, Jr.
Melvin W. Carter
George W. Casarett
Randall S. Caswell
Arthur B. Chilton
Stephan Cleary
Cyril L. Comar
Patricia Durbin
Merril Eisenbud
Thomas S. Ely
Benjamin G. Ferris
Asher J. Finkel
Donald C. Fleckenstein
Richard F. Foster
Hymer L. Friedell
Arthur H. Gladstein
Robert A. Goepp
Marvin Goldman
Robert O. Gorson
Arthur W. Guy
Ellis M. Hall
John H. Harley
Robert J. Hasterlik
John W. Healy
John M. Heslep
Marylou Ingram
Seymour Jablon
Jacob Kastner

Edward B. Lewis
Charles W. Mays
Roger O. McClellan
Mortimer Mendelsohn
Dade W. Moeller
Russell H. Morgan
Paul E. Morrow
Robert D. Moseley, Jr.
James V. Neel
Robert J. Nelsen
Peter C. Nowell
Frank Parker
Herbert M. Parker
Chester R. Richmond
Lester Rogers
Harald H. Rossi
Robert E. Rowland
William L. Russell
John H. Rust
Eugene L. Saenger
Harry F. Schulte
Raymond Selster
Warren K. Sinclair
Walter S. Snyder
Lewis V. Spencer
J. Newell Stannard
Chauncey Starr
John B. Storer
Lauriston S. Taylor
Roy C. Thompson
E. Dale Trout
Arthur C. Upton
John C. Villforth
George L. Voelz
Niel Wald
Edward W. Webster
George M. Wilkening
Harold O. Wyckoff

146 / THE NCRP

Honorary Members

Edgar C. Barnes	Paul C. Hodges
Carl B. Braestrup	George V. LeRoy
Austin M. Brues	Karl Z. Morgan
Frederick P. Cowan	Edith H. Quimby
Robley D. Evans	Shields Warren

Currently, the following Scientific Committees are actively engaged in formulating recommendations:

SC-1: Basic Radiation Protection Criteria
SC-7: Monitoring Methods and Instruments
SC-11: Incineration of Radioactive Waste
SC-18: Standards and Measurements of Radioactivity for Radiological Use
SC-23: Radiation Hazards Resulting from the Release of Radionuclides into the Environment
SC-24: Radionuclides and Labeled Organic Compounds Incorporated in Genetic Material
SC-25: Radiation Protection in the Use of Small Neutron Generators
SC-26: High Energy X-Ray Dosimetry
SC-28: Radiation Exposure from Consumer Products
SC-30: Physical and Biological Properties of Radionuclides
SC-31: Selected Occupational Exposure Problems Arising from Internal Emitters
SC-32: Administered Radioactivity
SC-33: Dose Calculations
SC-34: Maximum Permissible Concentrations for Occupational and Non-Occupational Exposures
SC-37: Procedures for the Management of Contaminated Persons
SC-38: Waste Disposal
SC-39: Microwaves
SC-40: Biological Aspects of Radiation Protection Criteria
SC-41: Radiation Resulting from Nuclear Power Generation
SC-42: Industrial Applications of X Rays and Sealed Sources
SC-44: Radiation Associated with Medical Examinations
SC-45: Radiation Received by Radiation Employees
SC-46: Operational Radiation Safety
SC-47: Instrumentation for the Determination of Dose Equivalent
SC-48: Apportionment of Radiation Exposure
SC-50: Surface Contamination
SC-51: Radiation Protection in Pediatric Radiology and Nuclear Medicine Applied to Children
SC-52: Conceptual Basis of Calculations of Dose Distributions
SC-53: Biological Effects and Exposure Criteria for Radiofrequency Electromagnetic Radiation
SC-54: Bioassay for Assessment of Control of Intake of Radionuclides

In recognition of its responsibility to facilitate and stimulate cooperation among organizations concerned with the scientific and related aspects of radiation protection and measurement, the Council has created a category of NCRP Collaborating Organizations. Organizations or groups of organizations which are national or international in scope and are concerned with scientific problems involving radiation quantities, units, measurements and effects, or radiation protection may be admitted to collaborating status by the Council. The present Collaborating Organizations with which the NCRP maintains liaison are as follows:

American Academy of Dermatology
American Association of Physicists in Medicine
American College of Radiology
American Dental Association
American Industrial Hygiene Association
American Insurance Association
American Medical Association
American Nuclear Society
American Occupational Medical Association
American Podiatry Association
American Public Health Association
American Radium Society
American Roentgen Ray Society
American Society of Radiologic Technologists
American Veterinary Medical Association
Association of University Radiologists
Atomic Industrial Forum
College of American Pathologists
Defense Civil Preparedness Agency
Genetics Society of America
Health Physics Society
National Bureau of Standards
National Electrical Manufacturers Association
Radiation Research Society
Radiological Society of North America
Society of Nuclear Medicine
United States Air Force
United States Army
United States Energy Research and Development Administration
United States Environmental Protection Agency
United States Navy
United States Nuclear Regulatory Commission
United States Public Health Service

The NCRP has found its relationships with these organizations to be extremely valuable to continued progress in its program.

The Council's activities are made possible by the voluntary contribution of the time and effort of its members and participants and the generous support of the following organizations:

Alfred P. Sloan Foundation
American Academy of Dental Radiology
American Academy of Dermatology
American Association of Physicists in Medicine
American College of Radiology
American College of Radiology Foundation
American Dental Association
American Industrial Hygiene Association
American Insurance Association
American Medical Association
American Mutual Insurance Alliance
American Nuclear Society
American Occupational Medical Association
American Osteopathic College of Radiology
American Podiatry Association
American Public Health Association
American Radium Society
American Roentgen Ray Society
American Society of Radiologic Technologists
American Veterinary Medical Association
American Veterinary Radiology Society
Association of University Radiologists
Atomic Industrial Forum
Battelle Memorial Institute
College of American Pathologists
Defense Civil Preparedness Agency
Edward Mallinckrodt, Jr. Foundation
Genetics Society of America
Health Physics Society
James Picker Foundation
National Association of Photographic Manufacturers
National Bureau of Standards
National Electrical Manufacturers Association
Radiation Research Society
Radiological Society of North America
Society of Nuclear Medicine
United States Energy Research and Development Administration
United States Environmental Protection Agency
United States Navy
United States Public Health Service

To all of these organizations the Council expresses its profound appreciation for their support.

Initial funds for publication of NCRP reports were provided by a grant from the James Picker Foundation and for this the Council wishes to express its deep appreciation.

The NCRP seeks to promulgate information and recommendations based on leading scientific judgment on matters of radiation protection and measurement and to foster cooperation among organizations concerned with these matters. These efforts are intended to serve the public interest and the Council welcomes comments and suggestions on its reports or activities from those interested in its work.

NCRP Publications

NCRP publications are distributed by the NCRP Publications' office. Information on prices and how to order may be obtained by directing an inquiry to:

> NCRP Publications
> P.O. Box 30175
> Washington, D.C. 20014

The extant publications are listed below.

Lauriston S. Taylor Lectures

No.	Title and Author
1	*The Squares of the Natural Numbers in Radiation Protection* by Herbert M. Parker
2	*Why be Quantitative About Radiation Risk Estimates?* by Sir Edward Pochin

NCRP Reports

No.	Title
8	*Control and Removal of Radioactive Contamination in Laboratories* (1951)
9	*Recommendations for Waste Disposal of Phosphorus-32 and Iodine-131 for Medical Users* (1951)
12	*Recommendations for the Disposal of Carbon-14 Wastes* (1953)
16	*Radioactive Waste Disposal in the Ocean* (1954)
22	*Maximum Permissible Body Burdens and Maximum Permissible Concentrations of Radionuclides in Air and in Water for Occupational Exposure* (1959) [Includes Addendum 1 issued in August 1963]
23	*Measurement of Neutron Flux and Spectra for Physical and Biological Applications* (1960)

25	*Measurement of Absorbed Dose of Neutrons and of Mixtures of Neutrons and Gamma Rays* (1961)
27	*Stopping Powers for Use with Cavity Chambers* (1961)
30	*Safe Handling of Radioactive Materials* (1964)
32	*Radiation Protection in Educational Institutions* (1966)
33	*Medical X-Ray and Gamma-Ray Protection for Energies Up to 10 MeV—Equipment Design and Use* (1968)
35	*Dental X-ray Protection* (1970)
36	*Radiation Protection in Veterinary Medicine* (1970)
37	*Precautions in the Management of Patients Who Have Received Therapeutic Amounts of Radionuclides* (1970)
38	*Protection against Neutron Radiation* (1971)
39	*Basic Radiation Protection Criteria* (1971)
40	*Protection Against Radiation From Brachytherapy Sources* (1972)
41	*Specification of Gamma-Ray Brachytherapy Sources* (1974)
42	*Radiological Factors Affecting Decision-Making in a Nuclear Attack* (1974)
43	*Review of the Current State of Radiation Protection Philosophy* (1975)
44	*Krypton-85 in the Atmosphere—Accumulation, Biological Significance, and Control Technology* (1975)
45	*Natural Background Radiation in the United States* (1975)
46	*Alpha-Emitting Particles in Lungs* (1975)
47	*Tritium Measurement Techniques* (1976)
48	*Radiation Protection for Medical and Allied Health Personnel* (1976)
49	*Structural Shielding Design and Evaluation for Medical Use of X Rays and Gamma Rays of Energies Up to 10 MeV* (1976)
50	*Environmental Radiation Measurements* (1976)
51	*Radiation Protection Design Guidelines for 0.1-100 MeV Particle Accelerator Facilities* (1977)
52	*Cesium-137 From the Environment to Man: Metabolism and Dose* (1977)
53	*Review of NCRP Radiation Dose Limit for Embryo and Fetus in Occupationally-Exposed Women* (1977)
54	*Medical Radiation Exposure of Pregnant and Potentially Pregnant Women* (1977)

152 / NCRP PUBLICATIONS

55 *Protection of the Thyroid Gland in the Event of Releases of Radioiodine* (1977)
56 *Radiation Exposure From Consumer Products and Miscellaneous Sources* (1977)
57 *Instrumentation and Monitoring Methods for Radiation Protection* (1978)
58 *A Handbook of Radioactivity Measurements Procedures* (1978)
59 *Operational Radiation Safety Program* (1978)
60 *Physical, Chemical, and Biological Properties of Radiocerium Relevant to Radiation Protection Guidelines* (1978)
61 *Radiation Safety Training Criteria for Industrial Radiography* (1978)
62 *Tritium in the Environment* (1978)

Binders for NCRP Reports are available. Two sizes make it possible to collect into small binders the "old series" of reports (NCRP Reports Nos. 8-31) and into large binders the more recent publications (NCRP Reports Nos. 32-62). Each binder will accommodate from five to seven reports. The binders carry the identification "NCRP Reports" and come with label holders which permit the user to attach labels showing the reports contained in each binder.

The following bound sets of NCRP Reports are also available:

> Volume I. NCRP Reports Nos. 8, 9, 12, 16, 22
> Volume II. NCRP Reports Nos. 23, 25, 27, 30
> Volume III. NCRP Reports Nos. 32, 33, 35, 36, 37
> Volume IV. NCRP Reports Nos. 38, 39, 40, 41
> Volume V. NCRP Reports Nos. 42, 43, 44, 45, 46
> Volume VI. NCRP Reports Nos. 47, 48, 49, 50, 51
> Volume VII. NCRP Reports Nos. 52, 53, 54, 55, 56, 57

(Titles of the individual reports contained in each volume are given above.)

The following NCRP reports are now superseded and/or out of print:

NCRP Report
No.   Title

1   *X-Ray Protection* (1931). [Superseded by NCRP Report No. 3]
2   *Radium Protection* (1934). [Superseded by NCRP Report No. 4]

3   *X-Ray Protection* (1936). [Superseded by NCRP Report No. 6]

4   *Radium Protection* (1938). [Superseded by NCRP Report No. 13]

5   *Safe Handling of Radioactive Luminous Compounds* (1941). [Out of print]

6   *Medical X-Ray Protection up to Two Million Volts* (1949). [Superseded by NCRP Report No. 18]

7   *Safe Handling of Radioactive Isotopes* (1949). [Superseded by NCRP Report No. 30]

10 *Radiological Monitoring Methods and Instruments* (1952). [Superseded by NCRP Report No. 57]

11 *Maximum Permissible Amounts of Radioisotopes in the Human Body and Maximum Permissible Concentrations in Air and Water* (1953). [Superseded by NCRP Report No. 22]

13 *Protection Against Radiations from Radium, Cobalt-60 and Cesium-137* (1954). [Superseded by NCRP Report No. 24]

14 *Protection Against Betatron—Synchrotron Radiations Up to 100 Million Electron Volts* (1954). [Superseded by NCRP Report No. 51]

15 *Safe Handling of Cadavers Containing Radioactive Isotopes* (1953). [Superseded by NCRP Report No. 21]

17 *Permissible Dose from External Sources of Ionizing Radiation* (1954) including *Maximum Permissible Exposure to Man, Addendum to National Bureau of Standards Handbook 59* (1958). [Superseded by NCRP Report No. 39]

18 *X-Ray Protection* (1955). [Superseded by NCRP Report No. 26]

19 *Regulation of Radiation Exposure by Legislative Means* (1955). [Out of print]

20 *Protection Against Neutron Radiation Up to 30 Million Electron Volts* (1957). [Superseded by NCRP Report No. 38]

21 *Safe Handling of Bodies Containing Radioactive Isotopes* (1958). [Superseded by NCRP Report No. 37]

24 *Protection Against Radiations from Sealed Gamma Sources* (1960). [Superseded by NCRP Reports Nos. 33, 34, and 40]

26 *Medical X-Ray Protection Up to Three Million Volts* (1961). [Superseded by NCRP Reports Nos. 33, 34, 35, and 36]

28 *A Manual of Radioactivity Procedures* (1961). [Superseded by NCRP Report No. 58]

29 *Exposure to Radiation in an Emergency* (1962). [Superseded by NCRP Report No. 42]

31 *Shielding for High Energy Electron Accelerator Installations* (1964). [Superseded by NCRP Report No. 51]

34 *Medical X-Ray and Gamma-Ray Protection for Energies Up to 10 MeV—Structural Shielding Design and Evaluation* (1970). [Superseded by NCRP Report No. 49]

Statements

The following statements of the NCRP were published outside of the NCRP Report series:

"Blood Counts, Statement of the National Committee on Radiation Protection," Radiology **63,** 428 (1954)

"Statements on Maximum Permissible Dose from Television Receivers and Maximum Permissible Dose to the Skin of the Whole Body," Am. J. Roentgenol., Radium Ther. and Nucl. Med. **84,** 152 (1960) and Radiology **75,** 122 (1960)

X-Ray Protection Standards for Home Television Receivers, Interum Statement of the National Council on Radiation Protection and Measurements (National Council on Radiation Protection and Measurements, Washington, 1968)

Specification of Units of Natural Uranium and Natural Thorium (National Council on Radiation Protection and Measurements, Washington, 1973)

Copies of the statements published in journals may be consulted in libraries. A limited number of copies of the last two statements listed above are available for distribution by NCRP Publications.

Index

Absorbed dose, D_1, 43
Accelerators, 1, 2, 13, 14, 18, 19, 32–35, 47, 100, 101, 128, 129
  Applications of accelerators, 2
  Accelerators of particles above 100 MeV, 2
  Betatrons, 2
  Cockcroft-Waltons, 2
  Cyclotrons, 2, 35, 128, 129
  Deuteron accelerators, 2
  Direct accelerators, 2, 33, 34, 35, 100
  Dynamitrons, 2
  Electron-beam generators, 2, 18, 19
  Electron linear accelerators, 2, 18, 19, 34
  Insulating-core transformers, 2
  Ion accelerators, 2
  Indirect accelerators, 2
  Linacs (see Electron linear accelerators)
  Neutron generators, 1, 2
  Pelletrons, 2
  Research accelerators, 2
  Synchrotrons, 2
  Tandem accelerators, 2, 33, 34, 35, 101
  Types of accelerators, 2
  Van de Graaff accelerators, 2, 14, 35
  X-ray generators for radiotherapy and radiography, 2, 13, 32, 47
Apertures in shielding barriers, 60, 65–68
  Mazes and ducts (see also Shielding thickness calculations: Apertures), 60
  Openings for material flow, 67
    pipes and ducts for continuous flow of homogeneous material, 67
    pipes for conveying samples, 68
    tunnels and ducts for intermittent flow of heterogeneous material, 67
  Shielded doors, 65
    materials, 65
    mounting and structural details, 65
    overlap and undercutting, 65
    thresholds and bottoms of doors, 66
    mechanical drives, 67

Area Occupancy Factor, U, 48, 91
  Nonoccupationally exposed persons, 91
  Occupationally exposed persons, 91
Areas, 11, 13, 15, 42
  Controlled, 11, 42
  Exclusion, 11, 13, 15
  High-radiation, 11
  Non controlled, 11
  Radiation, 11, 13, 15

Charged particles, 27–29, 93–95, 135
  Electrons, 27, 93, 94, 135
  Light ions (H and He), 28, 95
  Heavy ions (Li and heavier), 29
Conversion factors and equivalents, 92

Definitions, 73, 83
  Definitions of terms, 73
  Definitions of symbols, 83
Doors (see Apertures), 65
Dose equivalent index, H_I, 5, 43
  Equation for calculating H_I, 43
Dose limit, 3, 42, 85
  Dose limit for the public, 85
  Population dose limits, 85
  Maximum permissible does equivalent, 85
Ducts (see Apertures, also Shielding thickness calculations: Apertures), 61

Electron accelerators, electron-beam generators, 2, 18, 19, 34
Electrons, 27, 28, 93, 94, 135
  Range, 28, 93
  Scattering, 28, 94
  Collision stopping power in air, 135
Emission rates (see Neutron emission rates, Radiation emission rates, X-ray emission rates)
Exposure of individuals, 3, 85
  Maximum permissible dose equivalent, H_M, 3, 85
  Dose limits, 3, 85

Facility considerations, 7, 19
  Radiation safety systems, 7
  Siting and layout, 7
  Special problems, 19
Facility design basis, 1, 2, 4, 7, 8, 10, 19, 67, 68
  Architectural/engineering problems, 8, 10
  Beam dumps for x rays or neutrons, 10
  Horizontally oriented accelerators, 7
  Industrial process lines, 8, 19, 67, 68
  Information from accelerator manufacturers, 4
  Mobile or portable accelerators, 8
  Radiation emission rate, 4
  Research facilities, 2, 7
  Vertically mounted accelerators, 7
Fire hazards, 18
  from thermal heating effects, 18
  from radiation catalysis, 18
  from electrical effects of irradiation, 18
  interlocking with flow of material, 19

G-value, 134
  Production of O_3 by electron irradiation of O_2, 135
Gases, 17, 19, 20, 134, 136
  Insulating gases, e.g., N_2, CO_2, SF_6, 17
  Ozone, 19, 134
  Oxides of nitrogen, 19, 20, 134, 136
  Noxious gases, 17, 19, 134

Interlocks, 11–14, 19, 71
  By-passes, 12
  Control systems, 12, 14
  Electron-beam generators, 19
  Electrical hazards, 12
  Emergency switches, 12
  Fail-safe criterion, 12, 13
  Moveable radiation shielding, 12, 71
  Personnel access (entrance), 11
  Radiotherapy accelerators: electron vs x-ray operation, 13
  Redundancy, 12
  Resistance to radiation damage, 12

Ion sources and injectors, 35, 36
  Radiation hazards with cyclotrons, linacs, tandems, 35
  RF ion sources, 36
Ions, 28, 29, 95
  Light ions (H and He), 28
  range, 95
  deuterons, 29, 95
  Heavy ions (Li and heavier), 29

Klystrons, 36
  Shielding thickness, 36
  Radiation leakage, 36

Leakage radiation from radiotherapy accelerators, 36
  Protective tube housing, 36
  Dose-equivalent limitation, 36
  Dose equivalent from neutrons, 36

Material transfer systems, 22
  Pneumatic rabbits, 22
Maximum permissible dose equivalent, H_M, 3, 42, 85
Mazes (see Apertures), 61

Neutron emission rates, 101, 112, 114–116
  Angular distributions, 116
  Fluence rates or yields
    (p,n) reactions, 112
    (d,n) reactions, 114
    (γ,n) and (f,n) reactions, 115
  Neutrons from two-stage tandem accelerators, 101
Neutron energies from proton and deuteron reactions, 117
Neutron spectra, 120–122
  Accelerator conditions for generating neutron spectra, 120
  Transmission (dose-equivalent) through concrete:
    for (γ,n) and (f,n) reactions, 121
    for ion-induced reactions, 122
Nitrogen oxides, 19, 20, 134
Noxious gases, 17, 19, 134, 136
  Threshold limit values in workroom air, 136

Observation windows (see Visual observation of radiation areas), 16
Occupancy factor, U (see Area occupancy factor), 48, 91
Ozone, 19, 134, 136
  Production by external electron beams, 134
  Threshold limit values in workroom air, 136
  Concentration after accelerator shutoff, 136

INDEX / 157

Qualified expert, 5, 15
  Accelerator startup, 15
  Health physics, 5
  Radiation shielding, 5
  Radiological physics, 5
Quality factors, 55, 86, 87
  X rays and electrons, 86
  Neutrons, 87
  Variation in quality factor, 55

Radiation areas (see Areas)
Radiation damage, 17, 19, 23, 88
  Closed circuit television systems, 17
  Electron-beam generators: chemical damage from ozone, 19
  Insulating materials, 23
  Radiation paralysis of electronic devices, 23
  Recommendations for minimizing hazards, 23
  Thresholds for radiation damage: x rays, electrons, neutrons, 88
Radiation emission rates, 16, 26, 27
  (see also Neutron emission rates, X-ray emission rates)
  Angular distribution of radiation, 26
  Energy of accelerated particle, 26
  Energy spectrum of radiation, 27
  Indirect measure of emission rate, 16
  Radiation field map of accelerators, 27
  Species of accelerated particle, 26
  Target material, 26
Radiation monitoring systems, 15
  Fail-safe monitors, 15
  Indirect measurements of emission rate, 5, 15, 16
    electrical parameters of accelerators, back-up dosimeter or timer, 16
    neutron flux-density measurements, 16
    parallel-plate transmission ion chamber, 16
  Induced radioactivity, area radiation monitoring, 16
  Mixed fields, neutrons and x rays, 15
  Pulsed sources of radiation, 16
  Visual observation of radiation areas, 16
    closed-circuit television, 16
    periscopic arrangement through mazes, 16
    radiation shielded windows, 16
Radiation-protection surveillance, 5

Radiation sources, 16, 25–27, 29, 33–36
  Accelerator malfunction, 25
  Electrons, backscattered, 33
  General, 25–27, 29, 33, 35, 36
    accelerated charged particles, 27
    ion sources and injectors, 35
    leakage radiation from radiotherapy accelerators, 36
    other sources of radiation, 33
    penetrating radiations, 29
    radiation emission rates, 26
    radiation from accelerators, 25
  Induced radioactivity, 26
  Materials struck by direct x rays, 33
  Secondary electrons from direct accelerators, 33
    backstreaming electrons, 33
  Neutron sources (tandems), 34
  Spurious radiations, 34
    acceleration by capacitance, 35
    back-accelerated electrons from linacs, 35
    cyclotrons: ions lost in dee structure, 35
    poor vacuum, 34
    Van de Graaff frictional charge, 35
  Pulsed sources of radiations, 16
Radiations, penetrating, 29–32
  X rays, 29
    bremsstrahlung, 30
    characteristic, 30
    energies below 1.67 MeV, 31
    (γ,n) nuclear reactions, 30
    production, 30
  Neutrons, 31
    accelerators for radiotherapy and radiography, 32
    neutron-producing reactions, 31
    (d,n) reactions, 32
    (p,n) reactions, 32
    (γ,n) reactions, 32
    photoneutrons, 31
Radioactivity, 16, 20–22, 26, 32, 35, 41, 127–129
  Airborne radioactivity, 20
  Gamma-radiation following cyclotron shutdown, 129
  Good housekeeping, 21
  Handling, storage, disposal of radioactivity, 21
  Induced radioactivity, 16, 21, 26, 32, 35
  Induced radioactivity in cyclotrons, 128

Material transfer systems, 22
Natural radioactivity, 41
Photoneutron reactions, 127
Radioactive accelerator targets, 21
Radioactive components, 21
Radioactive contamination, 21
Radioactive products, 22
Tritium and other radioactive products in targets, 22
Water, 21
Reflection vs scattering, 9

Safety precautions, 17, 18
  Electrical circuits and interconnections, 17
  Fire hazards, 18
  First aid materials and equipment, 17
  Good housekeeping, 17
  Handling, conveying, storing materials, 17
  Special insulating gases, 17
  Toxic and noxious supplies and materials, 17
Safety systems, 11, 14, 15
  Accelerator operation, 14
    accelerator startup procedures, 15
    built-in protective systems, 14
    cyclotron RF excitation circuit, 14
    keyed switch, 14
    high-voltage power supply, 14
    radiation-producing circuits, 14
    Van de Graaff charging-belt drive, 14
  Interlocks, 11 (see Interlocks)
  Lights or audible signal system, 15
  Qualified individual in charge, 11
Scattering vs reflection, 9
Shielding, 37, 40, 42, 49
  Evaluation of shielding materials, 37
  Commonly used materials, 37
  Parameters of shielding calculations, 42
  Special purpose materials, 40
  Shielding thickness calculations, 49
Shielding, detachable, moveable, temporary, 12, 71, 72
  Barrier chains and stanchions, 72
  Direct accelerators of electrons and/or ions, 72
  Low-energy accelerators, ions, 71
  Moveable shielding, 12
  Radiotherapy x-ray generators, 72
Shielding materials, 27, 37–40, 130–132
  Evaluation of shielding materials, 37
  Commonly used materials, 37
  Materials for electrons, 37
  Materials for ions, 38
  Materials for neutrons, 39
    concrete, 39, 131, 132
    elemental composition of concrete, 131
    effect of water content and aggregate of concrete on transmission, 132
    materials with high absorption coefficient, 39
    miscellaneous materials, 133
    organic oils, paraffin, plastics, water, wood, 40
    very thick shielding barriers, 40
  Materials for x rays and gamma rays, 38, 130
    concrete, 38
    steel, 38
    lead, 38
  Shielding thickness calculations (see Shielding thickness calculations for individual radiations), 49
  Special purpose materials, 40
Shielding thickness calculations: Apertures, 60–62, 64, 65, 125
  Doors, 65
  Mazes and ducts, electrons, 61
  Mazes and ducts, neutrons, 62, 125
  Mazes and ducts, x rays, 61
  Neutron transmission factor through mazes or ducts, B_{nm}, 64, 125
Shielding thickness calculations: Charged particles, 49, 93, 95
  Electron ranges in air, water, tissue, aluminum, lead, 93
  Proton ranges for aluminum, copper, lead, uranium, 95
  Ranges for ^{2}H, ^{3}H, ^{3}He, ^{4}He, 95
Shielding thickness calculations: Computer programs, 72
Shielding thickness calculations: Multiple radiation sources, 59
  Direct accelerators: optional acceleration of ions or electrons, 59
  Complex accelerators with one or more injectors or accelerators in series, 59
Shielding thickness calculations: Neutrons, 55, 56, 58, 59, 87, 112, 114–116, 118, 121, 122, 125, 126

Shielding thickness calculations: Neutrons—*continued*
  Angular distribution of neutrons, 58, 116
  Fluence rate of neutrons, 112
  Generation of gamma rays in shield, 55
  Neutron shielding factor, B_n, 55
  Primary neutrons, 55
  Quality factors, 55, 87
  Reflected neutrons, 58
    general reflection within large facility, 58
    reflection from collimated beams, 58
    reflection coefficient, α_n, 59, 126
    thermal-neutron transmission through mazes and ducts, 125
  Transmission curves (dose equivalent), 56
    monoenergetic neutrons, 118
    spectra from (γ,n) and (γ,fn) reactions, 121
    spectra from ion-induced reactions, 122
  Yields of neutrons, 114, 115
Shielding thickness calculations: Skyshine, 68, 69, 71
  Neutron skyshine, 69
  Neutron skyshine transmission factor, B_{ns}, 71
  X-ray skyshine, 69
Shielding thickness calculations: X rays, 49, 50, 52–55, 103–111
  Equivalent electron energies for analysis of transmission of x rays emitted in 90° direction, 102
  Primary x rays, 50
  Reflected x rays, 53
  Reflected x rays from electron energies
    below 0.5 MeV, 54
    between 0.5 and 3 MeV, 54
    between 3 and 10 MeV, 55
    above 10 MeV, 55
  Reflection coefficient, α_x, 53, 111
  Shielding transmission factor, B_x, 49
  Reflected shielding transmission factor, B_{xr}, 54
  Tenth value layers: concrete, steel, lead, 108, 109, 110
  Transmission curves for broad-beam geometry: concrete, steel, lead, 103, 104, 105, 106, 107
  Transmission characteristics of reflected x rays, 53

X rays from externally produced electron beams, 52
Shielding thickness parameters used in calculations, 42–48
  Absorbed dose index, D_I, 43
  Area occupancy factor, U, 48
  Average hourly H_M, and dose limit, 42
  Average weekly H_M, and dose limit, 42
  Diagrams of radiation sources, 46, 47
  Dose equivalent index, H_I, 43
  Dose limits, 42
  General equation for calculating H_I, 43
  Geometrical factors, 45
    distances between radiation sources and occupiable areas, 45
    direction of accelerated particle beam, 45
    floor and ceiling barriers, 45
  Maximum permissible dose equivalent, H_M, 42
  Radiation source parameters, 44
    types of radiation, 44
    locations of radiations, 44
  Workload factor, W, 47
Siting and layout of facility, 7, 8
  Architectural/engineering problems, 8
  Industrial process lines, 8
  Modification of existing structures, 7
  Portable or mobile accelerators, 8
  Separate building or wing, 7
Space requirements in facility, 8–10, 12
  Accelerator room, servicing, 8
  Architectural/engineering problems, 10
  Auxiliary apparatus, 9
  Beam dumps for neutrons or x rays, 10
  Control room, 9, 12
  Experimental areas, 9
  Irradiation room, 9
  Low-level counting rooms, 10
  Photographic or radiographic darkrooms, 10
  Shops and offices, 10
Skyshine (see Shielding thickness calculations: Skyshine)

Toxic materials, 17
  Mercury-diffusion vacuum pumping systems, 17
Transmission of radiation, 102–110, 118, 120, 121, 124, 125
  Neutron transmission through concrete: monoenergetic neutrons, 118

neutron spectra from (γ,n) and (γ,fn) reactions, 120
neutron spectra from ion-induced reactions, 121
tenth-value layers in concrete, 120, 124
thermal-neutron transmission through mazes and ducts, 125
X-ray transmission:
transmission through concrete, steel, lead, 103–107
equivalent electron energies for analysis of transmission of x rays emitted in 90° direction, 102
tenth-value layers in concrete, steel, lead, 108–110
Tritium and tritons, 22, 29
 Gas in accelerators, 22, 29
 Inspection of accelerator components and system, 22
 Targets, 22, 29
 Tritium ions, 22

Ventilation and ducting, 19, 20, 60

Airborne radioactivity, 20
Filtration of air, 20
Production of ozone and other noxious gases for radiotherapy and electron accelerators, 19
Visual observation of radiation areas, 16
 Closed-circuit television, 16, 17
 Observation windows, 41
 Periscopic arrangements, 16
 Radiation damage to components, 17
 Radiation shielded windows, 16

Workload factor for radiotherapy installations, W, 47, 90

X-ray emission rates, 96, 98, 99–101
 Angular distribution from high-Z targets, 98
 X rays from backstreaming electrons in direct proton accelerators, 100
 X rays from low-Z targets, 99
 X rays from high-Z targets, 96
 X rays from two-stage tandem accelerators, 101